THE

HISTORY OF CREATION:

OR THE DEVELOPMENT OF THE EARTH AND ITS INHABITANTS BY THE ACTION OF NATURAL CAUSES.

A POPULAR EXPOSITION OF
THE DOCTRINE OF EVOLUTION IN GENERAL, AND OF THAT OF
DARWIN, GOETHE, AND LAMARCK IN PARTICULAR.

FROM THE GERMAN OF

ERNST HAECKEL,

IN TWO VOLUMES.
VOL. I.

Edited by Janice M. Hughes, Ph.D.

Foundations in Biological Thought No. 5

Briar Bird Press

First printing

The History of Creation: Or the Development of the Earth and its Inhabitants by the Action of Natural Causes, Volume I (1880) by Ernst Heinrich Haeckel available in the Public Domain

Edited material and additional material by Janice M. Hughes

Library and Archives Canada Cataloging in Publication

Hughes, Janice M. (Janice Maryan), 1958-, editor
The history of creation: or the development of the Earth and its inhabitants by the action of natural causes, volume 1 by Ernst Heinrich Haeckel / Foundations in biological thought, no. 5 edited by Janice M. Hughes

ISBN 978-0-9938707-6-7

Published in Canada by Briar Bird Press
Thunder Bay, Ontario, Canada
www.briarbirdpress.ca

Design and production by Janice M. Hughes
Cover design by Janice M. Hughes

Back cover photograph of Ernst Heinrich Haeckel circa 1874.

EDITOR'S PREFACE.

Ernst Heinrich Haeckel (1834–1919) was a German zoologist with expansive interests, including embryology, evolution, comparative anatomy, philosophy, and illustrative art; in addition, he achieved a medical degree in 1857. During his medical studies in Berlin, Haeckel was also introduced to marine biology by mentor Johannes Müller, a physiologist specializing in fish and other sea life. After two years of medical practice Haeckel, dissatisfied with his vocation, departed for Messina in southern Italy where he studied marine protozoans.

In 1859, Haeckel's life as a zoologist was solidified upon reading *On the Origin of Species by Means of Natural Selection* by Charles Darwin; he subsequently completed a dissertation in zoology at the University of Jena. Haeckel became a champion for the theory of evolution, and as a professor of comparative anatomy he published several monographs on the subject, many including his own intricately beautiful illustrations. Despite Haeckel's position of prestige among 19th-century scientists, his works were not without controversy. For example, his study of Radiolaria, protozoans with crystalline-like form, established his belief in spontaneous generation: that organic life can arise from inorganic matter. Moreover, his Biogenetic Law, which maintains that the development of an embryo (ontogeny) recapitulates the adult stages of organisms from which it evolved (phylogeny), has since been relegated to the realm of "biological mythology."

In *The History of Creation: Or the Development of the Earth and its Inhabitants by the Action of Natural Causes*, Haeckel presents his views on his peers and their doctrines, and his scientific perspectives of life on earth. The work was originally published as a two-volume set in 1880; however, it is often reprinted as a single abridged volume. This present edition is a return to the original two volume set; Volume II will follow as *Foundations in Biological Thought, no. 6*. In editing and reprinting this work, I have represented Haeckel's original manuscript in both appearance and intent through careful selection of fonts and formats. This edition also includes the original illustrations, footnotes, and scientific nomenclature. Page numbers and references have also been retained; page numbers specific to this edition have been attributed accordingly to maintain the utility of this book for modern readers.

Even today, Ernst Haeckel remains one of the most brilliant and controversial men of his time. As such, his *History of Creation* presents opinions that are often reverent, and sometimes scathing, which may be somehow more reflective of the author himself than the subjects of his critiques.

—Janice M. Hughes

Photograph of Ernst Heinrich Philipp August Haeckel taken by Nicola Perscheid in 1904.

THE HISTORY OF CREATION.

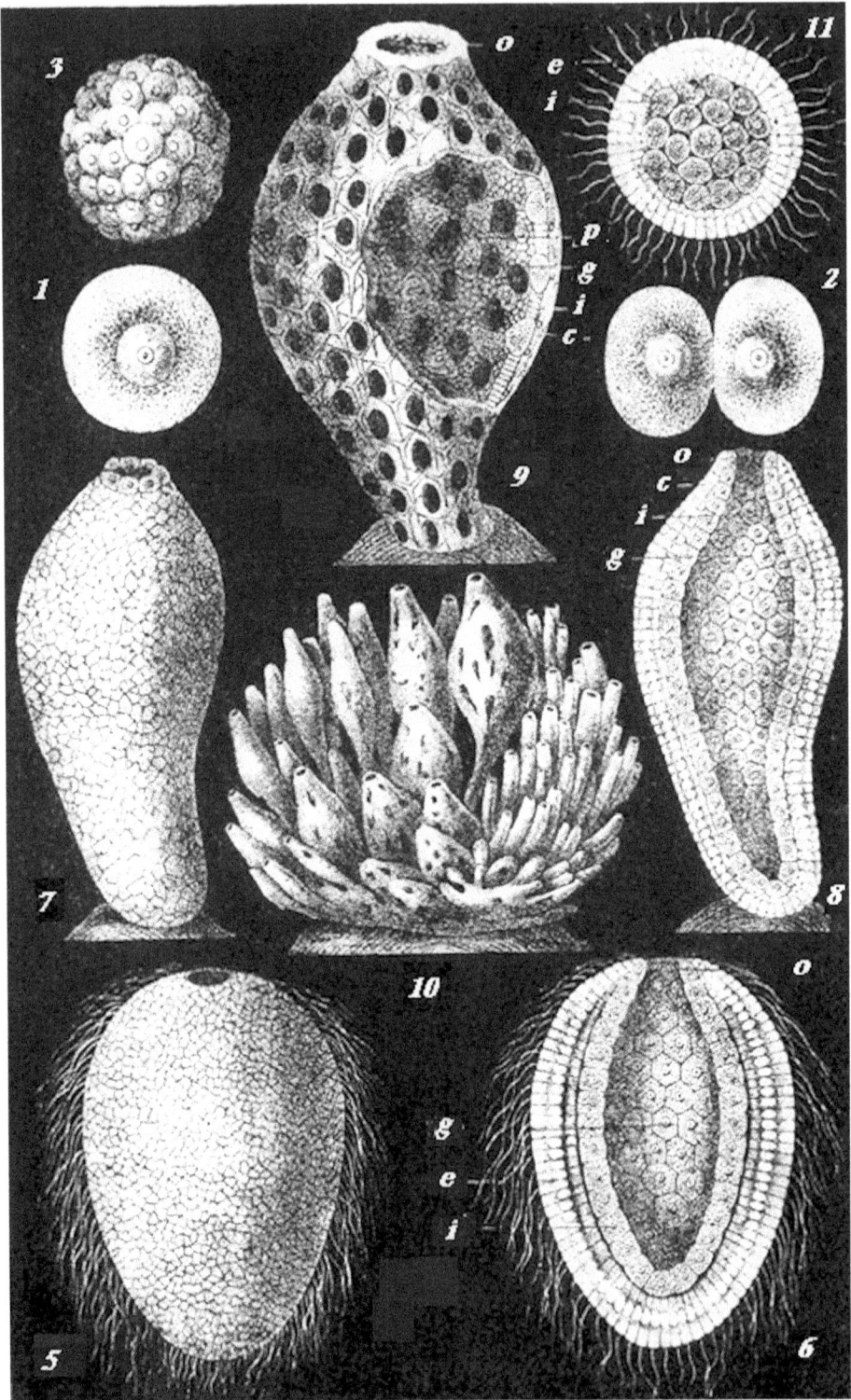

Development of a Calcareous Sponge (*Olynthus*).

THE

HISTORY OF CREATION:

OR THE DEVELOPMENT OF THE EARTH AND ITS
INHABITANTS BY THE ACTION OF NATURAL CAUSES.

A POPULAR EXPOSITION OF
THE DOCTRINE OF EVOLUTION IN GENERAL, AND OF THAT OF
DARWIN, GOETHE, AND LAMARCK IN PARTICULAR.

FROM THE GERMAN OF

ERNST HAECKEL,

PROFESSOR IN THE UNIVERSITY OF JENA.

THE TRANSLATION REVISED BY
E. RAY LANKESTER, M. A., FELLOW OF EXETER COLLEGE, OXFORD.

IN TWO VOLUMES.
VOL. I.

NEW YORK:
D. APPLETON AND COMPANY,
1, 3, AND 5 BOND STREET.
1880.

A sense sublime
Of something far more deeply interfused,
Whose dwelling is the light of setting suns,
And the round ocean, and the living air,
And the blue sky, and in the mind of man;
A motion and a spirit that impels
All thinking things, all objects of all thought,
And rolls through all things.

———————

In all things, in all natures, in the stars
Of azure heaven, the unenduring clouds,
In flower and tree, in every pebbly stone
That paves the brooks, the stationary rocks,
The moving waters and the invisible air.

WORDSWORTH.

CONTENTS OF VOL. I.*

CHAPTER III.

THE HISTORY OF CREATION ACCORDING TO CUVIER AND AGASSIZ.

CHAPTER IV.

THEORY OF DEVELOPMENT ACCORDING TO GOETHE AND OKEN.

CHAPTER V.

THEORY OF DEVELOPMENT ACCORDING TO KANT AND LAMARCK.

CHAPTER VI.

THEORY OF DEVELOPMENT ACCORDING TO LYELL AND DARWIN.

CHAPTER VII.

THE THEORY OF SELECTION (DARWINISM).

CHAPTER VIII.

TRANSMISSION BY INHERITANCE AND PROPAGATION.

CHAPTER IX.

LAWS OF TRANSMISSION BY INHERITANCE. ADAPTATION AND NUTRITION.

CHAPTER X.

LAWS OF ADAPTATION.

CHAPTER XI.

NATURAL SELECTION BY THE STRUGGLE FOR EXISTENCE. DIVISION OF LABOUR AND PROGRESS.

CHAPTER XII.

LAWS OF DEVELOPMENT OF ORGANIC TRIBES AND OF INDIVIDUALS. PHYLOGENY AND ONTOGENY.

CHAPTER XIII.

THEORY OF THE DEVELOPMENT OF THE UNIVERSE AND OF THE EARTH. SPONTANEOUS GENERATION. THE CARBON THEORY. THE PLASTID THEORY.

CHAPTER XIV.

MIGRATION AND DISTRIBUTION OF ORGANISMS. CHOROLOGY AND THE ICE-PERIOD OF THE EARTH.

LIST OF ILLUSTRATIONS.

PLATES.

FIGURES.

AUTHOR'S PREFACE TO THE ENGLISH EDITION.

I AM desirous of prefacing the English edition of the "History of Creation" with a few remarks which may serve to explain the origin and object of this book. In the year 1866 I published, under the title "Generelle Morphologie," a somewhat comprehensive work, which constituted the first attempt to apply the general doctrine of development to the whole range of organic morphology (Anatomy and Biogenesis), and thus to make use of the vast march onwards which the genius of Charles Darwin has effected in all biological science by his reform of the Descent Theory and its establishment through the doctrine of selection. At the same time, in the "Generelle Morphologie," the first attempt was made to introduce the Descent Theory into the systematic classification of animals and plants, and to found a "natural system" on the basis of genealogy; that is, to construct hypothetical pedigrees for the various species of organisms.

The "Generelle Morphologie" found but few readers, for which the voluminous and unpopular style of treatment, and its too extensive Greek terminology, may be chiefly to blame. But a proportionately large measure of approval has met the "Naturliche Schöpfungsgeschichte" in Germany. This book took its origin in the shorthand notes of a course of lectures which treated, before a mixed audience and in a popular form, the most important topics discussed in the "Generelle Morphologie." The notes were subsequently revised, and received considerable additions. The book appeared first in 1868, its fourth edition in 1873, and has been translated into several languages. I hope that it may also find sympathy in the fatherland of Darwin, the more so since it contains special morphological evidence in favour of many of the important doctrines with which this greatest naturalist of our century has enriched science. Proud as England may be to be called the fatherland of Newton, who, with his law of gravitation, brought inorganic nature under the dominion of natural laws of cause and effect, yet may she with even greater pride reckon Charles Darwin among her sons—he who solved the yet harder problem of bringing the complicated phenomena of organic nature under the sway of the same natural laws.

The reproach which is now oftenest made against the Descent Theory is that it is not securely founded, not sufficiently proven. Not only its distinct opponents maintain that there is a want of satisfactory proofs, but even faint-hearted and wavering adherents declare that Darwin's hypothesis is still wanting fundamental proof. Neither the former nor the latter estimate rightly the immeasurable weight which the great series of phenomena of comparative anatomy and ontogeny, palaeontology and taxonomy, chorology and oecology, cast into the scale in favour of the doctrine of filiation. Darwin's Theory of Selection, which completely explains the origin of species through the combined action of Inheritance and Adaptation in the struggle for existence, also appears to these persons not sufficient. They demand, over and above, that the descent of species from common ancestral forms shall be proved in a particular case; that, in contradistinction to the *synthetic* proofs adduced for the Descent Theory, the *analytic* proof of the genealogical continuity of the several species shall be brought forward.

This "analytical solution of the problem of the origin of species" I have myself endeavoured to afford in my recently published "Monograph of the Calcareous Sponges." For five consecutive years I have investigated this small but highly instructive group of animals in all its forms in the most careful manner, and I venture to maintain that the monograph, which is the result of those studies, is the most complete and accurate morphological analysis of an entire organic group which has up to this time been made. Provided with the whole of the material for study as yet brought together, and assisted by numerous contributions from all parts of the world, I was able to work over the whole group of organic forms known as the Calcareous Sponges in that greatest possible degree of fulness which appeared indispensable for the proof of the common origin of its species. This particular animal group is especially fitted for the analytical solution of the species problem, because it presents exceedingly simple conditions of organisation, because in it the morphological conditions possess a greatly superior, and the physiological conditions an inferior, import, and because all species of Calcispongiae are remarkable for the fluidity and plasticity of their form. With a view to these facts, I made two journeys to the sea-coast (1869 to Norway, 1871 to Dalmatia), in order to study as large a number of individuals as possible in their natural circumstances, and to collect specimens for comparison. Of many species, I compared several hundred individuals in the most careful way. I examined with the microscope and measured in the most accurate manner the details of form of all the species. As the final result of these exhaustive and almost endless examinations and measurements it appeared that "good species," in the ordinary dogmatic sense of the systematists, have no existence at all among the Calcareous Sponges;

that the most different forms are connected one with another by numberless gradational transition forms; and that all the different species of Calcareous Sponges are derived from a single exceedingly simple ancestral form, the *Olynthus*. A drawing of the *Olynthus* and its earliest stages of development (observe especially the highly important Gastrula) is given in the frontispiece of the present edition. Illustrations of the various structural details which establish the derivation of all Calcareous Sponges from the *Olynthus*, are given in the atlas of sixty plates which accompanies my monograph of the group. In the gastrula, moreover, is now also found the common ancestral form from which all the tribes of animals (the lowest group, that of the protozoa, alone being excepted) can without difficulty be derived. It is one of the most ancient and important ancestors of the human race!

If we take for the limitation of genus and species an average standard, derived from the actual practice of systematists, and apply this to the whole of the Calcareous Sponges at present known, we can distinguish about twenty-one genera, with one hundred and eleven species (as I have done in the second volume of the Monograph). I have, however, shown that we may draw up, in addition to this, another systematic arrangement (more nearly agreeing with the arrangement of the Calcispongiae hitherto in vogue) which gives thirty-nine genera and two hundred and eighty-nine species. A systematist who gives a more limited extension to the "ideal species" might arrange the same series of forms in forty-three genera and three hundred and eighty-one species, or even in one hundred and thirteen genera and five hundred and ninety species; another systematist on the other hand, who takes a wider limit for the "abstract species," would use in arranging the same series of forms only three genera, with twenty-one species, or might even satisfy himself with one genus and seven species. The delimitation of species and genera appears to be so arbitrary a matter, on account of endless varieties and transitional forms in this group, that their number is entirely left to the subjective taste of the individual systematist. In truth, from the point of view of the theory of descent, it appears altogether an unimportant question as to whether we give a wider or a narrower signification to allied groups of forms—whether we choose, that is to say, to call them genera or species, varieties or sub-species. The main fact remains undeniable, viz., the common origin of all the species from one ancestral form. The many-shaped Calcareous Sponges furnish, in the very remarkable conditions of their varieties of aggregation (metrocormy), a body of evidence in favour of this view which could hardly be more convincing. Not unfrequently the case occurs of several different forms growing out from a single "stock" or "cormus"—forms which until now have been regarded by systematists, not only as belonging to different species, but even to different genera. Fig. 10 in the frontispiece represents such a composite stock. This solid and tangible piece of evidence in favour of the common descent of different species ought, one would think, to satisfy the most determined sceptic!

In point of fact, I have a right to expect of my opponents that they shall carefully consider the "exact empirical proof" here brought forward for them, as they have so eagerly demanded. The opponents of the doctrine of filiation, who have too little power of weighing evidence, or possess too little knowledge to appreciate the overpowering weight of proof afforded by the synthetical argument (comparative anatomy, ontogeny, taxonomy, etc.), may yet be able to follow me along the path of analytical proof, and attempt to upset the conclusion as to the common origin of all species of all Calcareous Sponges which I have given in my Monograph. I must, however, repeat that this conclusion is based on the most minute investigation of an extraordinarily rich mass of material,—that it is securely established by thousands of the most careful microscopical observations, measurements, and comparisons of every single part, and that thousands of collected microscopic preparations render, at any moment, the most searching criticism of my results confirmatory of their correctness. One may hope, then, that opponents will endeavour to confront me on the ground of this "exact empiricism," instead of trying to damn my "nature-philosophical speculations." One may hope that they will endeavour to bring forward some evidence to show that the latter do not follow as the legitimate consequences of the former. May they, however, spare me the empty—though by even respectable naturalists oft-repeated—phrase, that the monistic nature-philosophy, as expounded in the "General Morphology," and in the "History of Creation," is wanting in actual proofs. The proofs are there. Of course those who turn their eyes away from them will not see them. Precisely that "exact" form of analytical proof which the opponents of the descent theory demand is to be found, by anybody who wishes to find it, in the "Monograph of the Calcareous Sponges."

Ernst Heinrich Haeckel.

Jena, June 24th, 1873.

NOTE.

FEELING sure that such a book as Professor Haeckel's "Schöpfungsgeschich-te" would do a great deal of good, if placed in the hands of the English read-ing public, and of commencing students of Natural History, I gladly under-took to revise for the publishers the present translation, which was made by a young lady. I have not attempted to escape a difficulty by ignoring the German names made use of by Professor Haeckel for classes, orders, and genera, but have adopted English equivalents. I do not submit these names as a maturely considered English nomenclature, they appear here simply as necessary parts of a close rendering of the German work. I do, however, hold that some such series of English terms is both possible and useful, and do not doubt—in spite of the pretended hostility of the genius of our language, and the curious sentimental objection that English names are *unscientific*—that we shall before long make use of plain English in speaking of the vari-ous groups of plants and animals—much to the gain of the larger public, and without detriment to the latinized nomenclature established for the purposes of the professional student.

E. R. L.

Oxford, October, 1874.

THE HISTORY OF CREATION.

CHAPTER I.

NATURE AND IMPORTANCE OF THE DOCTRINE OF FILIATION, OR DESCENT-THEORY.

General Importance and Essential Nature of the Theory of Descent as reformed by Darwin.—Its Special Importance to Biology (Zoology and Botany).—Its Special Importance to the History of the Natural Development of the Human Race.—The Theory of Descent as the Non-Miraculous History of Creation.—Idea of Creation.—Knowledge and Belief.—History of Creation and History of Development.—The Connection between the History of Individual and Palaeontological Development.—The Theory of Purposelessness, or the Science of Rudimentary Organs.—Useless and Superfluous Arrangements in Organisms.—Contrast between the two entirely opposed Views of Nature: the Monistic (mechanical, causal) and the Dualistic (teleological, vital).—Proof of the former by the Theory of Descent.—Unity of Organic and Inorganic Nature, and the Identity of the Active Causes in both.—The Importance of the Theory of Descent to the Monistic Conception of all Nature.

THE intellectual movement to which the impulse was given, thirteen years ago, by the English naturalist, Charles Darwin, in his celebrated work, "On the Origin of Species,"[1]* has, within this short period, assumed dimensions which cannot but excite the most universal interest. It is true the scientific theory set forth in that work, which is commonly called briefly Darwinism, is only a small fragment of a far more comprehensive doctrine—a part of the universal Theory of Development, which embraces in its vast range the whole domain of human knowledge.

But the manner in which Darwin has firmly established the latter by the former is so convincing, and the direction which has been given by the unavoidable conclusions of that theory to all our views of the universe, must appear to every thinking man of such deep significance, that its general

---

* Editor's Note: References are cited in full in volume II of *The History of Creation*.

importance cannot be over estimated. There is no doubt that this immense extension of our intellectual horizon must be looked upon as by far the most important, and rich in results, among all the numerous and grand advances which natural science has made in our day.

When our century, with justice, is called the age of natural science, when we look with pride upon the immensely important progress made in all its branches, we are generally in the habit of thinking more of immediate practical results, and less of the extension of our general knowledge of nature. We call to mind the complete reform, so infinitely rich in consequences to human intercourse, which has been effected by the development of machinery, by railways, steamships, telegraphs, and other inventions of physics. Or we think of the enormous influence which chemistry has brought to bear upon medicine, agriculture, and upon all arts and trades.

But much as we may value this influence of modern science upon practical life, still it must, estimated from a higher and more general point of view, stand most assuredly below the enormous influence which the theoretical progress of modern science will have on the entire range of human knowledge, on our conception of the universe, and on the perfecting of man's culture.

Think of the immense revolutions in all our theoretical views which we owe to the general application of the microscope. Think of the cell theory, which explains the apparent unity of the human organism as the combined result of the union of a mass of elementary vital units. Or consider the immense extension of our theoretical horizon which we owe to spectral analysis and to the mechanical theory of heat. But among all these wonderful theoretical advances, the theory wrought out by Darwin occupies by far the highest rank.

Every one of my readers has heard of the name of Darwin. But most persons have probably only an imperfect idea of the real value of his theory. If a reader estimates as of equal value all that has been written upon Darwin's memorable work since its appearance, the value of the theory will appear very doubtful to him, supposing that he has not been engaged in the organic natural sciences, and has not penetrated into the inner secrets of zoology and botany. The criticisms of it are so full of contradictions, and for the most part so defective, that we ought not to be at all astonished that even now, after the lapse of thirteen years since the appearance of Darwin's work, it has not gained half that importance which is justly due to it, and which sooner or later it certainly will attain.

Most of the innumerable writings which have been published during these years, both for and against Darwinism, are the productions of persons who are entirely wanting in the necessary amount of biological, and especially of zoological, knowledge. Although almost all of the more celebrated naturalists of the present day are adherents of the theory, yet only a few of them have endeavoured to procure its acceptance and recognition in larger circles. Hence the odd contradictions and the strange opinions which may still be heard everywhere about Darwinism. This is the reason which induces me to make Darwin's theory, and those further doctrines which are connected with it, the subject of these pages, which, I hope, will be generally intelligible. I hold it to be the duty of naturalists, not merely to meditate upon improvements and discoveries in the narrow circle to which their speciality confines them, not merely to pore over their one study with love and care, but also to seek to make the important general results of it fruitful to the mass, and to assist in spreading the knowledge of physical science among the people. The highest triumph of the human mind, the true knowledge of the most general laws of nature, ought not to remain the private possession of a privileged class of savans, but ought to become the common property of all mankind.

The theory which, through Darwin, has been placed at the head of all our knowledge of nature, is usually called the Doctrine of Filiation, or the Theory of Descent. Others term it the Transmutation Theory. Both designations are correct. For this doctrine affirms, that *all organisms* (viz., all species of animals, all species of plants, which have ever existed or still exist on the earth) *are derived from one single, or from a few simple original forms, and that they have developed themselves from these in the natural course of a gradual change.* Although this theory of development had already been brought forward and defended by several great naturalists, and especially by Lamarck and Goethe, in the beginning of our century, still it was through Darwin, thirteen years ago, that it received its complete demonstration and causal foundation; and this is the reason why now it is commonly and exclusively (though not quite correctly) designated as Darwin's Theory.

The great and really inestimable value of the Theory of Descent appears in a different light, accordingly as we merely consider its more immediate connection with organic natural science, or its larger influence upon the whole range of man's knowledge of the universe. Organic natural science, or Biology, which as Zoology treats of animals, as Botany of plants, is completely reformed and founded anew by the Theory of Descent. For by this theory we are made acquainted with the active causes of organic forms, while up to the present time Zoology and Botany have simply been occupied with the facts of these forms. We may therefore also term the theory of descent a *mechanical explanation of organic forms,* or the science of the true causes of Organic Nature.

As I cannot take for granted that my readers are all familiar with the terms "organic and inorganic nature," and as the contrast of both these natural bodies will, in future, occupy much of our attention, I must say a few words in explanation of them. We designate as *Organisms,* or *Organic bodies,* all *living creatures* or *animated bodies;* therefore all plants and animals, man included; for in them we can almost always prove a combination of various parts (instruments or organs) which work together for the purpose of producing the phenomena of life. Such a combination we do not find in *Anorgana*, or inorganic natural bodies—the so-called dead or *inanimate bodies*, such as minerals or stones, water, the atmospheric air, etc. Organisms always contain albuminous combinations of carbon in a semi-fluid condition of aggregation, which are always wanting in the Anorgana. Upon this important distinction rests the division of all natural history into two great and principal parts—*Biology*, or the science of Organisms (Zoology and Botany), and *Anorganology*, or the science of Anorgana (Mineralogy, Geology, Meteorology, etc.).

The great value of the Theory of Descent in regard to Biology consists, as I have already remarked, in its explaining to us the origin of organic forms in a mechanical way, and pointing out their active causes. But however highly and justly this service of the Theory of Descent may be valued, yet it is almost eclipsed by the immense importance which a single necessary inference from it claims for itself alone. This necessary and unavoidable inference is the theory of the *animal descent of the human race.*

The determination of the position of man in nature, and of his relations to the totality of things—this question of all questions for mankind, as Huxley justly calls it—is finally solved by the knowledge that man is descended from animals. In consequence of Darwin's reformed Theory of Descent, we are now in a position to establish scientifically the groundwork of a *non-miraculous history of the development of the human race.* All those who have defended Darwin's theory, as well as all its thoughtful opponents, have acknowledged that, as a matter of necessity, it follows from his theory that the human race, in the first place, must be traced to ape-like mammals, and further back to the lower vertebrate animals.

It is true Darwin himself did not express at first this most important of all the inferences from his theory. In his work, "On the Origin of Species," not a word is found about the animal descent of man. The courageous but cautious naturalist was at that time purposely silent on the subject, for he anticipated that this most important of all the conclusions of the Theory of Descent was at the same time the greatest obstacle to its being generally accepted and acknowledged. Certain it is that Darwin's book would have created, from the beginning, even much more opposition and offence, if this most important inference had at once been clearly expressed. It was not till twelve years later, in his work on "The Descent of Man, and Selection in Relation to Sex," that Darwin openly acknowledged that far-reaching conclusion, and expressly declared his entire agreement with those naturalists who had, in the meantime, themselves formed that conclusion. Manifestly the effect of this conclusion is immense, and *no* science will be able to escape from the consequences. Anthropology, or the science of man, and consequently all philosophy, are thereby thoroughly reformed in all their various branches.

It will be a later task in these pages to discuss this special point. I shall not treat of the theory of the animal descent of man till I have spoken of

Darwin's theory, and its general foundation and importance. To express it in one word, that most important, but (to most men) at first repulsive, conclusion is nothing more than a special deduction, which we must draw from the general inductive law of the descent theory (now firmly established), according to the stern commands of inexorable logic.

Perhaps nothing will make the full meaning of the theory of descent clearer than calling it the "*non-miraculous history of creation.*" I have therefore chosen that name for this work. It is, however, correct only in a certain sense, and it must be borne in mind that, strictly speaking, the expression "non-miraculous history of creation" contains a "*contradictio in adjecto.*"

In order to understand this, let us for a moment examine somewhat more closely what we understand by *creation*. If we understand the creation to mean the *coming into existence of a body* by a creative power or force, we may then either think of the *coming into existence of its substance* (corporeal matter), or of the *coming into existence of its form* (the corporeal form).

Creation in the former sense, as the *coming into existence of matter*, does not concern us here at all. This process, if indeed it ever took place, is completely beyond human comprehension, and can therefore never become a subject of scientific inquiry. Natural science teaches that matter is eternal and imperishable, for experience has never shown us that even the smallest particle of matter has come into existence or passed away. Where a natural body seems to disappear, as for example by burning, decaying, evaporation, etc., it merely changes its form, its physical composition or chemical combination. In like manner the coming into existence of a natural body, for example, of a crystal, a fungus, an infusorium, depends merely upon the different particles, which had before existed in a certain form or combination, assuming a new form or combination in consequence of changed conditions of existence. But never yet has an instance been observed of even the smallest particle of matter having vanished, or even of an atom being added to the already existing mass. Hence a naturalist can no more imagine the coming into existence of matter, than he can imagine its disappearance, and he therefore looks upon the existing quantity of matter in the universe as a given fact. If any person feels the necessity of conceiving the coming into existence of this matter as the work of a supernatural creative power, of the creative force of something outside of matter, we have nothing to say against it. But we must remark, that thereby not even the smallest advantage is gained for a scientific knowledge of nature. Such a conception of an immaterial force, which at the first creates matter, is an article of faith which has nothing whatever to do with human science. *Where faith commences, science ends.* Both these arts of the human mind must be strictly kept apart from each other. Faith has its origin in the poetic imagination; knowledge, on the other hand, originates in the reasoning intelligence of man. Science has to pluck the blessed fruits from the tree of knowledge, unconcerned whether these conquests trench upon the poetical imaginings of faith or not.

If, therefore, science makes the "history of creation" its highest, most difficult, and most comprehensive problem, it must accept as its idea of creation the second explanation of the word, viz., *the coming into being of the form* of natural bodies. In this way geology, which tries to investigate the origin of the inorganic surface of the earth as it now appears, and the manifold historical changes in the form of the solid crust of the earth, may be called the history of the creation of the earth. In like manner, the history of the development of animals and plants, which investigates the origin of living forms, and the manifold historical changes in animal and vegetable forms, may be termed the history of the creation of organisms. As, however, in the idea of creation, although used in this sense, the unscientific idea of a creator existing outside of matter, and changing it, may easily creep in, it will perhaps be better in future to substitute for it the more accurate term, *development*.

The great value which the *History of Development* possesses for the scientific understanding of animal and vegetable forms, has now been generally acknowledged for many years, and without it it would be impossible to make any sure progress in organic morphology, or the theory of forms. But by the history of development, only one part of this science has generally been understood, namely, that of organic individuals, usually called Embryology, but more correctly and comprehensively, *Ontogeny*. But, besides this, there is another history of development of organic species, genera, and tribes (phyla), which has the most important relations to the former.

The subject of this is furnished to us by the science of petrifactions, or palaeontology, which shows us that each tribe of animals and plants, during different periods of the earth's history, has been represented by a series of entirely different genera and species. Thus, for example, the tribe of vertebrated animals was represented by classes of fish, amphibious animals, reptiles, birds, and mammals, and each of these groups, at different periods, by quite different kinds. This palaeontological history of the development of organisms, which we may term *Phylogeny*, stands in the most important and remarkable relation to the other branch of organic history of development, I mean that of individuals, or Ontogeny. On the whole, the one runs parallel to the other. In fact, the history of individual development, or Ontogeny, is a short and quick recapitulation of palaeontological development, or Phylogeny, dependent on the laws of Inheritance and Adaptation.

As I shall have, later, to explain this most interesting and important coincidence more fully, I shall not dwell further upon it here, and merely call attention to the fact that it can only be explained and its causes understood by the Theory of Descent, while without that theory it remains completely incomprehensible and inexplicable. The Theory of Descent in the same way shows us *why* individual animals and plants must develop at all, and why they do not come into life at once in a perfect and developed state. No supernatural history of creation can in any way explain to us the great mystery of organic development. To this most weighty question, as well as to all other biological questions, the Theory of Descent gives us perfectly satisfactory answers—and always answers which refer to purely mechanical causes, and point to purely physico-chemical forces as the causes of phenomena which we were formerly accustomed to ascribe to the direct action of supernatural, creative forces. Hence, by our theory the mystic veil of the miraculous and supernatural, which has hitherto been allowed to hide the complicated phenomena of this branch of natural knowledge, is removed. All the departments of Botany and Zoology, and especially the most important portion of the latter, Anthropology, become reasonable. The dimming mirage of mythological fiction can no longer exist in the clear sunlight of scientific knowledge.

Of special interest among general biological phenomena are those which are quite irreconcilable with the usual supposition, that every organism is the product of a creative power, acting for a definite object. Nothing in this respect caused the earlier naturalists greater difficulty than the explanation of the so-called *"rudimentary organs,"*—those parts in animal and vegetable bodies which really have no function, which have no physiological importance, and yet exist in form. These parts deserve the most careful attention, although most unscientific men know little or nothing about them. Almost every organism, almost every animal and plant possesses, besides the obviously useful arrangements of its organization, other arrangements the purpose of which it is utterly impossible to make out.

Examples of this are found everywhere. In the embryos of many ruminating animals—among others, in our common cattle—fore-teeth, or incisors, are placed in the mid-bone of the upper jaw, which never fully develop, and therefore serve no purpose. The embryos of many whales—which afterwards possess the well-known whalebone instead of teeth—yet have before they are born, and while they take no nourishment, teeth in their jaws, which set of teeth never comes into use. Moreover, most of the higher animals possess muscles which are never employed; even man has such rudimentary muscles. Most of us are incapable of moving our ears as we wish, although the muscles for this movement exist, and although individual persons who have taken the trouble to exercise these muscles do succeed in moving their ears. It is still possible, by special exercise, by the persevering influence of the will upon the nervous system, to reanimate the almost extinct activity in the existing butimperfect organs, which are on the road to complete disappearance. On the other hand, we can no longer do this with another set of small rudimentary muscles, which still exist in the cartilage of the outer ear, but which are always perfectly inactive. Our long-eared ancestors of the tertiary period—apes, semi-apes, and pouched animals, like most other mammals, moved their large ear-flaps freely and actively; their muscles were much more strongly developed and of great importance. In a similar way, many varieties of dogs and rabbits, under the influence of civilized life, have left off "pricking up" their ears, and thereby have acquired imperfect auricular muscles and loose-hanging ears, although their wild ancestors moved their stiff ears in many ways.

Man has also these rudimentary organs on other parts of his body; they are of no importance to life, and never perform any function. One of the most remarkable, although the smallest organ of this kind, is the little crescent-like fold, the so-called "plica semilunaris," which we have in the inner corner of the eye, near the root of the nose. This insignificant fold of skin, which is quite useless to our eye, is the imperfect remnant of a third inner eyelid which, besides the upper and under eyelid, is highly developed in other mammals, and in birds and reptiles. Even our very remote ancestors of the Silurian period, the Primitive Fishes, seem to have possessed this third eyelid, the so-called nictitating membrane. For many of their nearest kin, who still exist in our day but little changed in form, viz., many sharks, possess a very strong nictitating membrane, which they can draw right across the whole eyeball, from the inner corner of the eye.

Eyes which do not see form the most striking example of rudimentary organs. These are found in very many animals, which live in the dark, as in caves or underground. Their eyes often exist in a well-developed condition, but they are covered by membrane, so that no ray of light can enter, and they can never see. Such eyes, without the function of sight, are found in several species of moles and mice which live underground, in serpents and lizards, in amphibious animals (*Proteus, Caecilia*), and in fishes; also in numerous invertebrate animals, which pass their lives in the dark, as do many beetles, crabs, snails, worms, etc.

An abundance of the most interesting examples of rudimentary organs is furnished by Comparative Osteology, or the study of the skeletons of vertebrate animals, one of the most attractive branches of Comparative Anatomy. In most of the vertebrate animals we find two pairs of limbs on the body, a pair of fore-legs and a pair of hind-legs. Very often, however, one or the other pair is imperfect; it is seldom that both are, as in the case of serpents and some varieties of eel-like fish. But some serpents, viz., the giant serpents (*Boa, Python*), have still in the hinder portion of the body some useless little bones, which are the remains of lost hind-legs.

In like manner the mammals of the whale tribe (Cetacea), which have only fore-legs fully developed (breast-fins,), have further back in their body another pair of utterly superfluous bones, which are remnants of undeveloped hind-legs. The same thing occurs in many genuine fishes, in which the hind-legs have in like manner been lost.

Again, in our slow-worm (*Anguis*), and in some other lizards, no fore-legs exist, although they have a perfect shoulder apparatus within their bodies, which should serve as a means of affixing the legs. Moreover, in various vertebrate animals, the single bones of both pairs of legs are found in all the different stages of imperfection, and often the degenerate bones and those muscles belonging to them are partially preserved, without their being able in any way to perform any function. The instrument is still there, but it can no longer play.

Moreover, we can, almost as generally, find rudimentary organs in the blossoms of plants, inasmuch as one part or another of the male organs of propagation—the stamen and anther, or of the female organs of propagation—the style, germ, etc.—is more or less imperfect or abortive. Among these we can trace, in various closely connected species of plants, the organ in all stages of degeneration. Thus, for example, the great natural family of lip-blossomed plants (Labiatae), to which the balm, peppermint, marjoram, ground-ivy, thyme, etc., belong, are distinguished by the fact that their mouth-like, two-lipped flower contains two long and two short stamens. But in many exceptional plants of this family, *e.g.* in different species of sage, and in the rosemary, only one pair of stamens is developed; the other pair is more or less imperfect, or has quite disappeared. Sometimes stamens exist, but without the anthers, so that they are utterly useless. Less frequently the rudiment or imperfect remnant of a fifth stamen is found, physiologically (for the functions of life) quite useless, but morphologically (for the knowledge of the form and of the natural relationship) a most valuable organ. In my "General Morphology of Organisms,"[4] in the chapter on "Purposelessness, or Dysteleology," I have given a great number of other examples (Gen. Morph. ii. 226).

No biological phenomenon has perhaps ever placed zoologists or botanists in greater embarrassment than these rudimentary or abortive organs. They are instruments without employment, parts of the body which exist without performing any service—adapted for a purpose, but without in reality fulfilling that purpose. When we consider the attempts which the earlier naturalists have made in order to explain this mystery, we can scarcely help smiling at the strange ideas to which they were led. Being unable to find a true explanation, they came, for example, to the conclusion that the Creator had placed these organs there "for the sake of symmetry," or they believed that it had appeared unwise and unsuitable to the Creator (seeing that their nearest kin did possess such organs) that these organs should be completely wanting in creatures, where they are incapable of performing a function, and where it cannot be otherwise from the special mode of life. In compensation for the non-existing function, he had at least furnished them with the outward but empty form; nearly in the same manner as civil officers, in uniform, are furnished with an innocent sword, which is never drawn from the scabbard. I scarcely believe, however, that any of my readers will be content with such an explanation.

Now, it is precisely this widely spread and mysterious phenomenon of rudimentary organs, in regard to which all other attempts at explanation fail, which is perfectly explained, and indeed in the simplest and clearest way, by Darwin's *Theory of Inheritance* and *Adaptation*. We can trace the important laws of inheritance and adaptation in the domestic animals which we breed, and the plants which we cultivate; and a series of such laws of inheritance have already been established. Without going further into this at present, I will only remark that some of them perfectly explain, in a mechanical way, the coming into existence of rudimentary organs, so that we must look upon the appearance of such structures as an entirely natural process, arising from the *disuse of the organs.*

By *adaptation* to special conditions of life, the formerly active and really working organs have gradually ceased to be used or employed. In consequence of their not being exercised they have become more and more imperfect, but in spite of this have always been handed down from one generation to another by *inheritance*, until at last they vanish partially or entirely. Now, if we admit that all the vertebrate animals mentioned above are derived from one common ancestor, possessing two seeing eyes and two well developed pairs of legs, the different stages of suppression and degeneration of these organs are easily accounted for in such of the descendants as could no longer use them. In like manner the various stages of suppression of the stamens, originally existing to the number of five (in the flower-bud), among the Labiatae is explained, if we admit that all the plants of this family sprung from one common ancestor, provided with five stamens.

I have here spoken somewhat fully of the phenomena of rudimentary organs, because they are of the utmost general importance, and because they lead us to the great, general, and fundamental questions in philosophy and natural science, for the solution of which the Theory of Descent has now become the indispensable guide. As soon, in fact, as, according to this theory, we acknowledge the exclusive activity of physico-chemical causes in living (organic) bodies, as well as in so-called inanimate (inorganic) nature, we concede exclusive dominion to that view of the universe, which we may designate as the *mechanical,* and which is opposed to the *teleological* conception. If we compare all the ideas of the universe prevalent among different nations at different times, we can divide them all into two sharply contrasted groups—a *causal* or *mechanical,* and a *teleological* or *vitalistic.* The latter has prevailed generally in Biology until now, and accordingly the animal and vegetable kingdoms have been considered as the products of a creative power, acting for a definite purpose. In the contemplation of every organism the unavoidable conviction seemed to press itself upon us, that such a wonderful machine, so complicated an apparatus for motion as exists in the organism, could only be produced by a power analogous to, but infinitely more perfect than, the power of man in the construction of his machines.

However sublime the former idea of a Creator, and his creative power, may have been; however much it may be attempted to divest it of all human analogy, yet in the end this analogy still remains unavoidable and necessary in the teleological conception of nature. In reality the Creator must himself be conceived of as an organism, that is, as a being who, analo-

themselves into the arms of a poetic faith, which as such can have no value in the domain of scientific knowledge.

All that was done before Darwin, to establish a natural mechanical conception of the origin of animals and plants, has been in vain, and until his time no theory gained a general recognition. Darwin's theory first succeeded in doing this, and thus has rendered an immense service. For the idea of the *unity of organic and inorganic nature* is now firmly established; and that branch of natural science which had longest and most obstinately opposed mechanical conception and explanation, viz., the science of the structure of animate forms, is launched on to identically the same road towards perfection as that along which all the rest of the natural sciences are travelling. The unity of *all* natural phenomena is by Darwin's theory finally established.

This unity of all nature, the animating of all matter, the inseparability of mental power and corporeal substance, Goethe has asserted in the words: "Matter can never exist and be active without mind, nor can mind without matter." These first principles of the mechanical conception of the universe have been taught by the great monistic philosophers of all ages. Even Democritus of Abdera, the immortal founder of the Atomic theory, clearly expressed them about 500 years before Christ; but the great Dominican friar, Giordano Bruno, did so even more explicitly. For this he was burnt at the stake, by the Christian inquisition in Rome, on the 17th of Feb., 1600, on the same day on which, 36 years before, Galileo, his great fellow-countryman and fellow-worker, was born. Such men, who live and die for a great idea, are usually stigmatized as "materialists"; but their opponents, whose arguments were torture and the stake, are praised as "spiritualists."

By the Theory of Descent we are for the first time enabled to conceive of the unity of nature in such a manner that a mechanico-causal explanation of even the most intricate organic phenomena, for example, the origin and structure of the organs of sense, is no more difficult (in a general way) than is the mechanical explanation of any physical process; as, for example, earthquakes, the courses of the wind, or the currents of the ocean. We thus arrive at the extremely important conviction that *all natural bodies* which are known to us are *equally animated*, that the distinction which has been made between animate and inanimate bodies does *not* exist. When a stone is thrown into the air, and falls to earth according to definite laws, or when in a solution of salt a crystal is formed, the phenomenon is neither more nor less a mechanical manifestation of life than the growth and flowering of plants, than the propagation of animals or the activity of their senses, than the perception or the formation of thought in man. This final triumph of the monistic conception of nature constitutes the highest and most general merit of the Theory of Descent, as reformed by Darwin.

CHAPTER II.

SCIENTIFIC JUSTIFICATION OF THE THEORY OF DESCENT. HISTORY OF CREATION ACCORDING TO LINNAEUS.

The Theory of Descent, or Doctrine of Filiation, as the Monistic Explanation of Organic Natural Phenomena.—Its Comparison with Newton's Theory of Gravitation.—Limits of Scientific Explanation and of Human Knowledge in general.—All Knowledge founded originally on Sensuous Experience, *à posteriori.*—Transition of *à posteriori* knowledge, by Inheritance, into *à priori* knowledge.—Contrast between the Supernatural Hypotheses of the Creation according to Linnaeus, Cuvier, Agassiz, and the Natural Theories of Development according to Lamarck, Goethe, and Darwin.—Connection of the former with the Monistic (mechanical), of the latter with the Dualistic Conception of the Universe.— Monism and Materialism.—Scientific and Moral Materialism.—The History of Creation according to Moses.—Linnaeus as the Founder of the Systematic Description of Nature and Distinction of Species.—Linnaeus' Classification and Binary Nomenclature.— Meaning of Linnaeus' Idea of Species.—His History of Creation.—Linnaeus' view of the Origin of Species.

THE value which every scientific theory possesses is measured by the number and importance of the objects which can be explained by it, as well as by the simplicity and universality of the causes which are employed in it as grounds of explanation. On the one hand, the greater the number and the more important the meaning of the phenomena explained by the theory, and the simpler, on the other hand, and the more general the causes which the theory assigns as explanations, the greater is its scientific value, the more safely we are guided by it, and the more strongly are we bound to adopt it.

Let us call to mind, for example, that theory which has ranked up to the present time as the greatest achievement of the human mind—the Theory of Gravitation, which Newton, two hundred years ago, established in his Mathematical Principles of Natural Philosophy. Here we find that the object to be explained is as large as one can well imagine. He undertook to reduce the phenomena of the motion of the planets, and the structure of the universe, to mathematical laws. As the most simple cause of these intricate phenomena of motion, Newton established the law of weight or attraction, the same law which is the cause of the fall of bodies, of adhesion, cohesion, and many other phenomena.

If we apply the same standard of valuation to Darwin's theory, we must arrive at the conclusion that this theory, also, is one of the greatest achievements of the human mind, and that it may be placed quite on a level with Newton's Theory of Gravitation. Perhaps this opinion will seem a little exaggerated, or at any rate very bold, but I hope in the course of this treatise to convince the reader that this estimate is not too high. In the preceding chapter, some of the most important and most general phenomena in organic nature, which have been explained by Darwin's theory, have been named. Among them are the variations in form which accompany the individual development of organisms, most varied and complicated phenomena, which until now presented the greatest difficulties in the way of mechanical explanation, that is, in the tracing of them to active causes. We have mentioned the *rudimentary organs*, those exceedingly remarkable structures in animals and plants which have no object and refute every teleological explanation seeking for the final purpose of the organism. A great number of other phenomena might have been mentioned, which are no less important, and are explained in the simplest manner by Darwin's reformed Theory of Descent. For the present I will only mention the phenomena presented to us by the *geographical distribution of animals and plants* on the surface of our planet, as well as the *geological distribution of the extinct and petrified organisms* in the different strata of the earth's crust. These important palaeontological and geographical phenomena, which were formerly only known to us as *facts*, are now traced to their active *causes* by the Theory of Descent.

The same statement applies further to all the general laws of *Comparative Anatomy*, especially to the great law of *division of labour* or *separation* (polymorphism, or differentiation), a law which determines the form or structure of human society, as well as the organization of individual animals and plants. It is this law which necessitates an ever *increasing variety*, as well as a *progressive development* of organic forms. This law of the division of labour has, up to the present time, been only recognized as a fact, and it, like the law of progressive development, or the law of progress which we perceive active everywhere in the history of nations (as also in that of animals and plants), is explained by Darwin's Doctrine of Descent. Then, if we turn our attention to the great whole of organic nature, if we compare all the individual groups of phenomena of this immense domain of life, it cannot fail to appear, in the light of the Doctrine of Descent, no longer as the ingeniously designed work of a Creator building up according to a definite purpose, but as the necessary consequence of active causes, which are inherent in the chemical combination of matter itself, and in its physical properties.

In fact, we can most positively assert, and I shall justify this assertion in the course of these pages, that by the Doctrine of Filiation, or Descent, we are enabled for the first time to reduce all organic phenomena to a single law, and to discover a single active cause for the infinitely intricate mechanism of

the whole of this rich world of phenomena. In this respect, Darwin's theory
stands quite on a level with Newton's Theory of Gravitation; indeed, it even
rises higher than Newton's theory!

The grounds of explanation are equally simple in the two theories. In
explaining this most intricate world of phenomena, Darwin does not make
use of new or hitherto unknown properties of matter, nor does he, as one
might suppose, make use of discoveries of new combinations of matter or of
new forces of organization; but it is simply by extremely ingenious combination, by the synthetic comprehension, and by the thoughtful comparison of
a number of well-known facts, that Darwin has solved the "holy mystery" of
the living world of forms. The consideration of the interchanging relations
which exist between two general properties of organisms, viz., *Inheritance
and Adaptation*, is what has here been of the first importance. Merely by
considering the relations between these two vital actions or physiological
functions of organisms, also further by considering the reciprocal interaction which all animals and plants, living in one and the same place, necessarily exert on one another—solely by the correct estimate of these simple
facts, and by skilfully combining them, Darwin has succeeded in finding the
true active causes (causae efficientes) of the immensely intricate world of
forms in organic nature.

In any case we are in duty bound to accept this theory till a better one
be found, which will undertake to explain the same amount of facts in an
equally simple manner. Until now we have been in utter want of such a
theory. The fundamental idea that all different animal and vegetable forms
must be descended from a few or even from one single, most simple primary
form, was indeed not new. This idea was long since distinctly formulated—
first by the great Lamarck, at the beginning of our century. But Lamarck in
reality only expressed the hypothesis of the Doctrine of Filiation, without
establishing it by an explanation of the active causes. And it is just the demonstration of these causes which marks the extraordinary progress which
Darwin's theory has made beyond that of Lamarck. In the physiological
properties of Inheritance and Adaptation of organic matter, Darwin discovered the true cause of the genealogical relationship of organisms. It was
not possible for the genius of Lamarck in his day to command that colossal
material of biological facts which has been collected by the patient zoological and botanical investigations of the last fifty years, and which has been
used by Darwin as an overpowering apparatus of evidence.

Darwin's theory is therefore not what his opponents frequently represent
it as being—an unwarranted hypothesis taken up at random. It is not for zoologists or botanists to accept or reject this as an explanatory theory, as they
please; they are rather compelled and obliged to accept it, according to the
general principle observed in all natural sciences, that we must accept and
retain for the explanation of phenomena any theory which, though it has
only a feeble basis, is compatible with the actual facts—until it is replaced

by a better one. If we do not adopt it, we renounce a scientific explanation of phenomena, and this is, in fact, the position which many biologists still maintain. They look upon the whole domain of animate nature as a perfect mystery, and upon the origin of animals and plants, the phenomena of their development and affinities, as quite inexplicable and miraculous; in fact, they will not allow that there *can* be a true understanding of them.

Those opponents of Darwin who do not exactly wish to renounce a scientific explanation are in the habit of saying, "Darwin's theory of the common origin of the different species is only *one* hypothesis; we oppose to it *another*, the hypothesis that the individual animal and vegetable species have not developed one from another by descent, but that they have come into existence independently of one another, by a still undiscovered law of nature." But as long as it is not shown how this coming into existence is to be conceived of, and what that "law of nature" is—as long as not even *probable* grounds of explanation can be brought forward to account for the independent coming into existence of animal and vegetable species, so long this counter-hypothesis is in fact no hypothesis, but an empty unmeaning phrase. Darwin's theory ought, moreover, not to be called an hypothesis. For a scientific hypothesis is a supposition, postulating the existence of unknown properties or motional phenomena of natural bodies, which properties have not as yet been observed by the experience of the senses. But Darwin's theory does not assume such unknown conditions; it is based upon general properties of organisms that have long been recognized, and—as has been remarked—it is the exceedingly ingenious and comprehensive combination of a number of phenomena which had hitherto stood isolated, which gives the theory its extraordinarily great and intrinsic value. By it we are for the first time in a position to demonstrate an active cause for all the known morphological phenomena in the animal and vegetable kingdoms; and, in fact, this cause is always one and the same, viz., the alternate action of Adaptation and Inheritance, therefore a physiological, that is, a physico-chemical or mechanical, relationship. For these reasons the acceptance of the Doctrine of Filiation, as mechanically established by Darwin, is a binding and unavoidable necessity for the whole domain of zoology and botany.

As, therefore, in my opinion the immense importance of Darwin's theory lies in the fact that it has *mechanically explained those organic phenomena of forms* which had hitherto been unexplained, it is perhaps necessary that I should here say a few words about the different ideas connected with the word "explanation." It is very frequently said, in opposition to Darwin's theory, that it does indeed explain those phenomena by Inheritance and Adaptation, but that it does not at the same time explain those properties of organic matter, and that therefore we do not arrive at first causes. This objection is quite correct, but it applies equally to *all* explanations of phenomena. We *nowhere* arrive at a knowledge of first causes. The origin of every simple salt crystal, which we obtain by evaporating its mother liquor, is no less mysterious to us, as far as concerns its first cause, and in itself no less incomprehensible than the origin of every animal which is developed out of a simple cell. In explaining the most simple physical or chemical phenomena, as the falling of a stone, or the formation of a chemical combination, we arrive, by discovering and establishing the active causes—for example, the gravitation or the chemical affinity—at other remoter phenomena, which in themselves are mysterious. This arises from the limitation or relativity of our powers of understanding. We must not forget that human knowledge is absolutely limited, and possesses only a relative extension. It is, in its essence, limited by the very nature of our senses and of our brains.

All knowledge springs from sensuous perceptions. In opposition to this statement, the innate, *à priori* knowledge of man may be brought up; but we can see that the so-called *à priori* knowledge can by Darwin's theory be proved to have been acquired *à posteriori*, being based on experience as its first cause. Knowledge which is based originally upon purely empirical observations, and which is therefore a purely sensuous experience, but has then been transmitted from generation to generation by inheritance, appears in later generations as if it were independent, innate, and *à priori*. In our late animal ancestors, all our so-called "*à priori knowledge*" was originally acquired *à posteriori*, and only gradually became *à priori* by inheritance. It is based in the first instance upon experiences, and by the laws of Inheritance and Adaptation we can positively prove that knowledge *à priori* and knowledge *à posteriori* cannot rightly be placed in opposition, as is usually done. On the contrary, sensuous experience is the original source of *all* knowledge. For this reason alone, all our knowledge is limited, and we can never apprehend the *first causes* of any phenomena. The force of crystallization, the force of gravitation, and chemical affinity remain in themselves just as incomprehensible as do Adaptation and Inheritance.

Seeing that Darwin's theory explains from a single point of view the totality of all those phenomena of which we have given a brief survey, that it demonstrates one and the same quality of the organism as the active cause in all cases, we must allow that it gives us for the present *all* that we can desire. Moreover, we have good reason to hope that at some future time we shall learn to explain the first causes at which Darwin has arrived, namely, the properties of Adaptation and Inheritance; and that we shall succeed in discovering in the composition of albuminous matter certain molecular relations as the remoter, simpler causes of these phenomena. There is indeed no prospect of this in the immediate future, and we content ourselves for the present with the tracing back of organic phenomena to two mysterious properties, just as in the case of Newton's theory we are satisfied with tracing the planetary motions to the force of gravitation, which itself is likewise a mystery to us and not cognizable in itself.

Before commencing our principal task, which is the careful discussion of the Doctrine of Descent, and the consequences that arise out of it, let us take an historical retrospect of the most important and most widely spread of those views, which before Darwin men had elaborated concerning organic creation, and the coming into existence of the many animal and vegetable species. In doing this I have no intention of entertaining the reader with a statement of all the innumerable stories about the creation which have been current among the different human species, races, or tribes. However interesting and gratifying this task would be, from an ethnographical point of view, as well as in a history of civilization, it would lead us here much too far from our subject. Besides, the great majority of all these legends about creation bear too clearly the stamp of arbitrary fiction, and of a want of a close observance of nature, to be of interest in a scientific treatment of the history of creation. I shall therefore only select the Mosaic history from among those that are not founded on scientific investigation, on account of the unparalleled influence which it has gained in the western civilized world; and then I shall immediately take up the scientific hypothesis about creation, which originated with Linnaeus as late as the commencement of last century.

All the different conceptions which man has ever formed about the coming into existence of the different animal and vegetable species may conveniently be divided into two great contrasted groups—the natural and supernatural histories of creation.

These two groups, on the whole, correspond with the two different principal forms of the human notions of the universe which we have already contrasted as the *monistic and the dualistic* conception of nature. In the usual dualistic or teleological (vital) conception of the universe, organic nature is regarded as the purposely executed production of a Creator working according to a definite plan. Its adherents see in every individual species of animal and plant an "embodied creative thought," the material expression of a *definite first cause* (causa finalis) acting for a set purpose. They must necessarily assume supernatural (not mechanical) processes for the origin of organisms. With justice, we may therefore designate their scheme of the world's growth as the *Supernatural History of Creation.* Among all such teleological histories of creation, that of Moses has gained the greatest influence, since even so distinguished a naturalist as Linnaeus has claimed admittance for it in Natural Science. Cuvier's and Agassiz's views of creation also belong to this group, as do in fact those of the great majority of both scientific and unscientific men.

On the other hand, the theory of development carried out by Darwin, which we shall have to treat of here as the *Non-miraculous* or *Natural History of Creation*, and which has already been put forward by Goethe and Lamarck, must, if carried out logically, lead to the monistic or mechanical (causal) conception of the universe. In opposition to the dualistic or teleological conception of nature, our theory considers organic, as well as inorganic, bodies to be the necessary products of natural forces. It does not see in every individual species of animal and plant the embodied thought of a personal Creator, but the expression for the time being of a mechanical process of development of matter, the expression of a necessarily active cause, that is, of a mechanical cause (causa efficiens). Where teleological Dualism seeks the arbitrary thoughts of a capricious Creator in the miracles of creation, causal Monism finds in the process of development the necessary effects of eternal immutable laws of nature.

The Monism here maintained by us is often considered identical with Materialism. Now, as Darwinism, and in fact the whole theory of development, has been designated as "*materialistic*," I cannot avoid here at once guarding myself against this ambiguous word, and against the malice with which, in certain quarters, it is employed to stigmatize our doctrine.

By the word "*Materialism*," two completely different things are very frequently confounded and mixed up, which in reality have nothing whatever to do with each other, namely, scientific and moral materialism. Scientific materialism, which is identical with our Monism, affirms in reality no more than that everything in the world goes on naturally—that every effect has its cause, and every cause its effect. It therefore assigns to causal law—that is, the law of a necessary connection between cause and effect—its place over the entire series of phenomena that can be known. At the same time, scientific materialism positively rejects every belief in the miraculous, and every conception, in whatever form it appears, of supernatural processes. Accordingly, nowhere in the whole domain of human knowledge does it recognize real metaphysics, but throughout only physics; through it the inseparable connection between matter, form, and force becomes self evident. This scientific materialism has long since been so universally acknowledged in the wide domain of inorganic science, in Physics and Chemistry, in Mineralogy and Geology, that no one now doubts its sole authority. But in Biology, or Organic science, the case is very different; here its value is still continually a matter of dispute in many quarters. There is, however, nothing else which can be set up against it, excepting the metaphysical spectre of a vital power, or empty theological dogma. If we can prove that all nature, so far as it can be known, is only *one*, that the same "great, eternal, iron laws" are active in the life of animals and plants, as in the growth of crystals and in the force of steam, we may with reason maintain the monistic or mechanical view of things throughout the domain of Biology—in Zoology and Botany—whether it be stigmatized as "materialism" or not. In such a sense all exact science, and the law of cause and effect at its head, is purely materialistic.

Moral, or *ethical Materialism*, is something quite distinct from scientific materialism, and has nothing whatever in common with the latter. This real materialism proposes no other aim to man in the course of his life than the most refined possible gratification of his senses. It is based on the delusion that purely material enjoyment can alone give satisfaction to man; but as he can find that satisfaction in no one form of sensuous pleasure, he dashes on weariedly from one to another. The profound truth that the real value of life does not lie in material enjoyment, but in moral action—that true happiness does not depend upon external possessions, but only in a virtuous course of life—this is unknown to ethical materialism. We therefore look in vain for such materialism among naturalists and philosophers, whose highest happiness is the intellectual enjoyment of Nature, and whose highest aim is the knowledge of her laws. We find it in the palaces of ecclesiastical princes, and in those hypocrites who, under the outward mask of a pious worship of God, solely aim at hierarchical tyranny over, and material spoliation of, their fellow-men. Blind to the infinite grandeur of the so-called "raw material," and the glorious world of phenomena arising from it—insensible to the inexhaustible charms of Nature, and without a knowledge of her laws— they stigmatize all natural science, and the culture arising from it, as sinful "materialism," while really it is this which they themselves exhibit in a most shocking form. Satisfactory proofs of this are furnished, not only by the whole history of the Catholic Popes, with their long series of crimes, but also by the history of the morals of orthodoxy in every form of religion.

In order, then, to avoid in future the usual confusion of this utterly objectionable Moral Materialism with our Scientific Materialism, we think it necessary to call the latter either *Monism* or *Realism*. The principle of this *Monism* is the same as what Kant terms the "principle of mechanism," and of which he expressly asserts, that *without it there can be no natural science at all.* This principle is quite inseparable from our Non-miraculous History of Creation, and characterizes it as opposed to the teleological belief in the miracles of a Supernatural History of Creation.

Let us now first of all glance at the most important of all the supernatural histories of creation, I mean that of Moses, as it has been handed down to us in the Bible, the ancient document of the history and laws of the Jewish people. The Mosaic history of creation, since in the first chapter of Genesis it forms the introduction to the Old Testament, has enjoyed, down to the present day, general recognition in the whole Jewish and Christian world of civilization. Its extraordinary success is explained not only by its close connection with Jewish and Christian doctrines, but also by the simple and natural chain of ideas which runs through it, and which contrasts favourably with the confused mythology of creation current among most of the other ancient nations. First the Lord God creates the earth as an inorganic body; then he separates light from darkness, then water from the dry land. Now the earth has become inhabitable for organisms, and plants are first created, animals later—and among the latter the inhabitants of the water and the air first, afterwards the inhabitants of the dry land. Finally God creates man, the last of all organisms, in his own image, and as the ruler of the earth.

Two great and fundamental ideas, common also to the non-miraculous theory of development, meet us in this Mosaic hypothesis of creation, with surprising clearness and simplicity—the idea of separation or *differentiation*, and the idea of progressive development or *perfecting*. Although Moses looks upon the results of the great laws of organic development (which we shall later point out as the necessary conclusions of the Doctrine of Descent) as the direct actions of a constructing Creator, yet in his theory there lies hidden the ruling idea of a progressive development and a differentiation of the originally simple matter. We can therefore bestow our just and sincere admiration on the Jewish lawgiver's grand insight into nature, and his simple and natural hypothesis of creation, without discovering in it a so-called "divine revelation." That it cannot be such is clear from the fact that two great fundamental errors are asserted in it, namely, first, the *geocentric* error that the earth is the fixed central point of the whole universe, round which the sun, moon, and stars move; and secondly, the *anthropocentric* error, that man is the premeditated aim of the creation of the earth, for whose service alone all the rest of nature is said to have been created. The former of these errors was demolished by Copernicus' System of the Universe in the beginning of the 16th century, the latter by Lamarck's Doctrine of Descent in the beginning of the 19th century.

Although the geocentric error of the Mosaic history was demonstrated by Copernicus, and thereby its authority as an absolutely perfect divine revelation was destroyed, yet it has maintained, down to the present day, such influence, that it forms in many wide circles the principle obstacle to the adoption of a natural theory of development. Even in our century, many naturalists, especially geologists, have tried to bring the Mosaic theory into harmony with the recent results of natural science, and have, for example, interpreted Moses' seven days of creation as seven great geological periods. However, all these ingenious attempts at interpretation have so utterly failed, that they require no refutation here. The Bible is no scientific book, but consists of records of the history, the laws, and the religion of the Jewish people, the high merit of which, as a history of civilization, is not impaired by the fact that in all scientific questions it has no commanding importance, and is full of gross errors.

We may now make a great stride over more than three thousand years, from Moses, who died about the year 1480 before Christ, to Linnaeus, who was born in the year 1707 after Christ. During this whole period no history of creation was brought forward that gained any lasting importance, or the closer examination of which would here be of any interest. Indeed, during the last fifteen hundred years, since Christianity gained its supremacy, the Mosaic history of creation, together with the dogmas connected with it, has become so generally predominant, that the 19th century is the first that has dared positively to rise against it. Even the great Swedish naturalist, Linnaeus, the founder of modern natural history, linked his System of Nature most closely to the Mosaic history of creation.

The extraordinary progress which Charles Linnaeus made in the so-called descriptive natural sciences, consists, as is well known, in his having established a system of nomenclature of animals and plants, which he carried out in a manner so perfectly logical and consistent, that down to the present day it has remained in many respects the standard for all succeeding naturalists engaged in the study of the forms of animals and plants. Although Linnaeus' system was artificial, although in classifying animal and vegetable species he only sought and employed single parts as the foundation for his divisions, it has, nevertheless, gained the greatest success; firstly, in consequence of its being carried out consistently, and secondly, by its nomenclature of natural bodies, which has become extremely important, and at which we must here briefly glance.

Before Linnaeus' time, many vain attempts had been made to throw light upon the endless chaos of different animal and vegetable forms (then known) by adopting for them suitable names and groupings; but Linnaeus, by a happy hit, succeeded in accomplishing this important and difficult task, when he established the so-called "*binary nomenclature.*" The binary nomenclature, or the twofold designation, as Linnaeus first established it, is still universally applied by all zoologists and botanists, and will, no doubt, maintain itself, for a long time to come, with undiminished authority. It consists in this, that every species of animal and plant is designated by two names, which stand to each other in the same relation as do the christian and surnames of a man. The special name which corresponds with the christian name, and expresses the idea of "a species," serves as the common designation of all individual animals or plants, which are equal in all essential matters of form, and are only distinguished by quite subordinate features. The more general name, on the other hand, corresponding with the surname, and which expresses the idea of a genus, serves for the common designation of all the most nearly similar kinds or species.

According to Linnaeus' plan, the more general and comprehensive generic name is written first; the special subordinate name of the species follows it. Thus, for example, the common cat is called *Felis domestica*; the wild cat, *Felis catus*; the panther, *Felis pardus*; the jaguar, *Felis onca*; the tiger, *Felis tigris*; the lion, *Felis leo*. All these six kinds of animals of prey are different species of one and the same genus—*Felis*. Or, to add an example from the vegetable kingdom, according to Linnaeus' designation the pine is *Pinus abies*; the fir, *Pinus picea*; the larch, *Pinus larix*; the Italian pine, *Pinus pinea*; the Siberian stone pine, *Pinus cembra*; the knee timber, *Pinus mughus*; the common pine, *Pinus silvestris*. All these seven kinds of pines are different species of one and the same genus—*Pinus*.

Perhaps this advance made by Linnaeus may seem to some only of subordinate importance in the practical distinction and designation of the variously formed organisms. But in reality it was of the very greatest importance, both from a practical and theoretical point of view. For now, for the first time, it became possible to arrange the immense mass of different organic forms according to their greater or less degree of resemblance, and to obtain an easy survey of the general outlines of such a "system." Linnaeus facilitated the tabulation and survey of this "system" of plants and animals still more by placing together the most nearly similar genera into so-called orders (ordines); and by uniting the most nearly similar orders into still more comprehensive main divisions or classes. Thus, according to Linnaeus, each of the two organic kingdoms were broken up into a number of classes, the vegetable kingdom into twenty-four, and the animal kingdom into six. Each class again contains several orders. Every single order may contain a number of genera, and, again, every single genus several species.

Valuable as was Linnaeus' binary nomenclature in a *practical* way, in bringing about a comprehensive systematic distinction, designation, arrangement, and division of the organic world of forms, yet the incalculable *theoretical* influence which it gained forthwith in relation to the history of creation was no less important. Even now all the important fundamental questions as to the history of creation turn finally upon the decision of the very remote and unimportant question, *What really are kinds or species?* Even now the *idea of organic species* may be termed the central point of the whole question of creation, the disputed centre, about the different conceptions of which Darwinists and Anti-Darwinists fight.

According to Darwin's opinion, and that of his adherents, the different species of one and the same genus of animals and plants are nothing else than differently developed descendants of one and the same original primary form. The different kinds of pine mentioned above would accordingly have originated from a single primaeval form of pine. In like manner the origin of all the species of cat mentioned above would be traced to a single common form of *Felis*, the ancestor of the whole genus. But further, in accordance with the Doctrine of Descent, all the different genera of one and the same order ought also to be descended from one common primary ancestor, and so, in like manner, all orders of a class from a single primary form.

On the other hand, according to the idea of Darwin's opponents, all species of animals and plants are quite independent of each other, and only the individuals of each species have originated from a single primary form. But if we ask them how they conceive these original primary forms of each species to have come into existence, they answer with a leap into the incomprehensible, "They were created."

Linnaeus himself defined the idea of species in this manner by saying, "There are as many different species as there were different forms created in the beginning by the infinite Being." ("Species tot sunt diversae, quot diversas formas ab initio creavit infinitum ens.") In this respect, therefore, he follows most closely the Mosaic history of creation, which in the same way maintains that animals and plants were created "each one after its kind." Linnaeus, accepting this, held that originally of each species of animals and plants either a single individual or a pair had been created; in fact a pair, or, as Moses says, "a male and a female" of those species which have separate sexes, but of those species in which each individual combines both sexual organs (hermaphrodites), as for instance the earthworm, the garden and vineyard snails, as well as the great majority of plants, a single individual.

Linnaeus further follows the Mosaic legend in regard to the flood, by supposing that the great general flood destroyed all existing organisms, except those few individuals of each species (seven pairs of the birds and of clean animals, one pair of unclean animals) which Noah saved in the ark, and which were placed again on land, on Mount Ararat, after the flood had subsided. He tried to explain the geographical difficulty of the living together of the most different animals and plants, as follows: Mount Ararat, in Armenia, being situated in a warm climate, and rising over 16,000 feet in height, combines in itself the conditions for a temporary common abode of such animals as live in different zones. Accordingly, animals accustomed to the polar regions could climb up the cold mountain ridges, those accustomed to a warm climate could go down to the foot of the mountain, and the inhabitants of a temperate zone could remain midway up the mountain. From this point it was possible for them to spread north and south over the earth.

It is scarcely necessary to remark that this Linnaean hypothesis of creation, which evidently was intended to harmonize most closely with the prevailing belief in the Bible, requires no serious refutation. When we consider Linnaeus' clearness and sagacity in other matters, we may doubt whether he believed it himself. As to the simultaneous origin of all individuals of each species from one pair of ancestors respectively (or in the case of the hermaphrodite species, from one original hermaphrodite), it is clearly quite untenable; for, apart from other reasons, in the first days after the creation, the few animals of prey would have sufficed to have utterly demolished all the herbivorous animals, as the herbivorous animals must have destroyed the few individuals of the different species of plants. The existence of such an equilibrium in the economy of nature as obtains at present cannot possibly be conceived, if only one individual of each species, or only one pair, had originally and simultaneously been created.

Moreover, how little importance Linnaeus himself attached to this untenable hypothesis of creation is clear, among other things, from the fact that he recognized *Hybridism* (crossing) as a source of the production of new species. He assumed that a great number of independent new species had originated by the interbreeding of two different species. Indeed, such hybrids are not at all rare in nature, and it is now proved that a great number of species, for example, of the genus *Rubus* (bramble), mullen (*Verbascum*), willow (*Salix*), thistle (*Cirsium*), are hybrids of different species of these genera. We also know of hybrids between hares and rabbits (two species of the genus *Lepus*), further of hybrids between different species of dog (genus *Canis*), etc., which can be propagated as independent species.

It is certainly very remarkable that Linnaeus asserted the physiological (therefore mechanical) origin of new species in this process of hybridism. It clearly stands in direct opposition to the supernatural origin of the other species by creation, which he accepted as put forward in the Mosaic account. The one set of species would therefore have originated by dualistic (teleological) creation, the other by monistic (mechanical) development.

The great and well merited authority which Linnaeus gained by his systematic classification and by his other services to Biology, was clearly the reason why his views of creation also remained, throughout the whole of the last century, undisputed and generally recognized. If throughout systematic Zoology and Botany the distinctions, classification, and designations of species, introduced by Linnaeus, and the dogmatic ideas connected therewith had not been maintained—more or less unaltered—we should be at a loss to understand how his idea of an independent creation of single species could have stood, by itself down to the present day. It is only owing to his great authority, and through his attaching himself to the prevailing Biblical belief, that his hypothesis of creation has retained its position so long.

CHAPTER III.

THE HISTORY OF CREATION ACCORDING TO CUVIER AND AGASSIZ.

General Theoretical Meaning of the Idea of Species.—Distinction between the Theoretical and Practical Definition of the Idea of Species.—Cuvier's Definition of Species.—Merits of Cuvier as the Founder of Comparative Anatomy.—Distinction of the Four Principal Forms (types or branches) of the Animal Kingdom, by Cuvier and Bär.—Cuvier's Services to Palaeontology.—His Hypothesis of the Revolutions of our Globe, and the Epochs of Creation separated by them.—Unknown Supernatural Causes of the Revolutions, and the subsequent New Creations.—Agassiz's Teleological System of Nature.—His Conception of the Plan of Creation, and its six Categories (groups in classification).—Agassiz's Views of the Creation of Species.—Rude Conception of the Creator as a man-like being in Agassiz's Hypothesis of Creation.—Its internal Inconsistency and Contradictions with the important Palaeontological Laws discovered by Agassiz.

THE real matter of dissension in the contest carried on by naturalists as to the origin of organisms, their creation and development, lies in the conceptions which are entertained about the *nature of species*. Naturalists either agree with Linnaeus, and look upon the different species as distinct forms of creation, independent of one another, or they assume with Darwin their blood-relationship. If we share Linnaeus' view (which was discussed in our last chapter), that the different organic species came into existence independently—that they have no blood-relationship—we are forced to admit that they were created independently, and we must either suppose that every single organic individual was a special act of creation (to which surely no naturalist will agree), or we must derive all individuals of every species from a single individual, or from a single pair, which did not arise in a natural manner, but was called into being by command of a Creator. In so doing, however, we turn aside from the safe domain of a rational knowledge of nature, and take refuge in the mythological belief in miracles.

If, on the other hand, with Darwin, we refer the similarity of form of the different species to real blood-relationship, we must consider all the different species of animals and plants as the altered descendants of one or a few most simple original forms. Viewed in this way, the Natural System of organisms (that is, their tree-like and branching arrangement and division into classes, orders, families, genera, and species) acquires the significance of a real genealogical tree, whose root is formed by those original archaic forms which have long since disappeared. But a truly natural and consistent view of organisms can assume no supernatural act of creation for even those simplest original forms, but only a coming into existence by *spontaneous generation** (archigony, or generatio spontanea). From Darwin's view of the nature of species, we arrive therefore at a *natural theory of development*; but from Linnaeus' conception of the idea of species, we must assume a *supernatural dogma of creation.*

Most naturalists after Linnaeus, whose great services in systematic and descriptive natural history won for him such high authority, followed in his footsteps, and without further inquiry into the origin of organization, they assumed, in the sense of Linnaeus, an independent creation of individual species, in conformity with the Mosaic account of creation. The foundation of their conception was based upon Linnaeus' words: "There are as many different species as there were different forms created in the beginning by the Infinite Being." We must here remark at once, without going further into the definition of species, that all zoologists and botanists in their classificatory systems, in the practical distinction and designation of species of animals and plants, never troubled, or even could trouble, themselves in the slightest degree about this assumed creation of the parent forms. In reference to this, one of our first zoologists, the ingenious Fritz Müller, makes the following striking observation: "Just as in Christian countries there is a catechism of morals, which every one knows by heart, but which no one considers it his duty to follow, or expects to see followed by others,—so zoology also has its dogmas, which are just as generally professed as they are denied in practice." (Für Darwin, p. 71.)[16]

Linnaeus' venerated dogma of species is just such an irrational dogma, and for that very reason it is powerful. Although most naturalists blindly submitted to it, yet they were, of course, never in a position to demonstrate the descent of individuals belonging to one species from the common, originally created, primitive form. Zoologists and botanists, in their systems of nomenclature, confined themselves entirely to the similarity of forms, in order to distinguish and name the different species. They placed in one species all organic individuals which were very similar, or almost identical in form, and which could only be distinguished from one another by very unimportant differences. On the other hand, they considered as different species those individuals which presented more essential or more striking differences in the formation of their bodies. But of course this opened the flood-gates to the most arbitrary proceedings in the systematic distinctions of species. For as all the individuals of one species are never completely alike in all their parts, but as every species varies more or less, no one could

---

* Archbiosis (Bastion), Abiogenesis (Huxley)

point out which degree of variation constituted a really "good species," or which degree indicated a "mere variety."

This dogmatic conception of the idea of species, and the arbitrary proceedings connected with it, necessarily led to the most perplexing contradictions, and to the most untenable suppositions. This is clearly demonstrable in the case of the celebrated Cuvier (born in 1769), who next to Linnaeus has exercised the greatest influence on the study of zoology. In his conception and definition of the idea of species, he agreed on the whole with Linnaeus, and shared also his belief in an independent creation of individual species. Cuvier considered their immutability of such importance that he was led to the foolish assertion—"The immutability of species is a necessary condition of the existence of scientific natural history." As Linnaeus' definition of species did not satisfy him, he made an attempt to give a more exact and, for systematic practice, a more useful definition, in the following words: "All those individual animals and plants belong to one species which can be proved to be either descended from one another, or from common ancestors, or which are as similar to these as the latter are among themselves."

In dealing with this matter, Cuvier reasoned in the following manner:— "In those organic individuals, of which we know that they are descended from one and the same common form of ancestors—in which, therefore, their common ancestry is empirically proved—there can be no doubt that they belong to one species, whether they differ much or little from one another, or whether they are almost alike or very unlike. Moreover, all those individuals also belong to this species which differ no more from the latter (those proved to be derived from a common stock) than these differ from one another." In a closer examination of this definition of species given by Cuvier, it becomes at once evident that it is neither theoretically satisfactory nor practically applicable. Cuvier, with this definition, began to move in the same circle in which almost all subsequent definitions of species have moved, through the assumption of their immutability.

Considering the extraordinary authority which George Cuvier has gained in the science of organic nature, and in consequence of the almost unlimited supremacy which his views exercised in zoology, during the first half of our century, it seems appropriate here to examine his influence a little more closely. This is all the more necessary as we have to combat, in Cuvier, the most formidable opponent to the Theory of Descent and the monistic conception of nature.

One of the many and great merits of Cuvier is that he stands forth as the founder of Comparative Anatomy. While Linnaeus established the distinction of species, genera, orders, and classes mostly upon external characters, and upon separate and easily discoverable signs in the number, size, place, and form of individual organic parts of the body, Cuvier penetrated much more deeply into the essence of organization. He demonstrated great and wide differences in the inner structure of animals, as the real foundation of a scientific knowledge and classification of them. He distinguished natural families in the classes of animals, and established his natural system of the animal kingdom on their comparative anatomy.

The progress from Linnaeus' artificial system to Cuvier's natural system was exceedingly important. Linnaeus had arranged all animals in a single series, which he divided into six classes, two classes of Invertebrate, and four classes of Vertebrate animals. He distinguished these artificially, according to the nature of their blood and heart. Cuvier, on the other hand, showed that in the animal kingdom there were four great natural divisions to be distinguished, which he termed Principal Forms, or General Plans, or Branches of the animal kingdom (Embranchments), namely—1. The Vertebrate animals (Vertebrata); 2. The Articulate animals (Articulata); 3. The Molluscous animals (Mollusca); and 4. The Radiate animals (Radiata). He further demonstrated that in each of these four branches a peculiar plan of structure or type was discernible, distinguishing each branch from the three others. In the Vertebrate animals it is distinctly expressed by the form of the skeleton, or bony framework, as also by the structure and position of the dorsal nerve-chord, apart from many other peculiarities. The Articulate animals are characterized by their ventral nerve-chord and their dorsal heart. In Molluscs the sack-shaped and non-articulate body is the distinguishing feature. The Radiate animals, finally, differ from the three other principal forms by their body being the combination of four or more main sections united in the form of radii (antimera).

The distinction of these four principal forms of animals, which has become extremely productive in the development of zoology, is commonly ascribed entirely to Cuvier. However, the same thought was expressed almost simultaneously, and independently of Cuvier, by Bär, one of the greatest naturalists, and still living, who did the most eminent service in the study of animal development. Bär showed that in the development of animals, also, four different main forms (or types) must be distinguished.[20] These correspond with the four plans of structure in animals, which Cuvier distinguished on the ground of comparative anatomy. Thus, for example, the individual development of all Vertebrate animals agrees, from the commencement, so much in its fundamental features that the germs or embryos of different Vertebrate animals (for example, of reptiles, birds, and mammals) in their earlier stages cannot be distinguished at all. It is only at a late stage of development that there gradually appear the more marked differences of form which separate those different classes and orders from one another. The plan of structure, which shows itself in the individual development of Articulate animals (insects, spiders, crabs), is from the beginning essentially the same in all Articulate animals, but different from that of all Vertebrate animals. The same holds good, with certain limitations, in Molluscous and Radiated animals.

Neither Bär, who arrived at the distinction of the four animal types or principal forms through the history of the individual development (Embryology), nor Cuvier, who arrived at the same conclusion by means of comparative anatomy, recognized the true cause of this difference. This is disclosed to us by the Theory of Descent. The wonderful and astonishing similarity in the inner organization and in the anatomical relations of structure, and the still more remarkable agreement in the embryonic development of all animals belonging to one and the same type (for example, to the branch of the Vertebrate animals), is explained in the simplest manner by the supposition of their common descent from a single primary original form. If this view is not accepted, then the complete agreement of the most different Vertebrate animals, in their inner structure and their manner of development, remains perfectly inexplicable. In fact it can only be explained by the law of *inheritance*.

Next to the comparative anatomy of animals and the systematic zoology founded anew by it, it was specially to the science of petrifactions, or Palaeontology, that Cuvier rendered great service. We must draw special attention to this, because these very palaeontological views, and the geological ideas connected with them, were held almost universally in the highest esteem during the first half of the present century, and caused the greatest hindrance to the working out of a truly natural history of creation.

Petrifactions, the scientific study of which Cuvier promoted at the beginning of our century in a most extensive manner, and established quite anew for the Vertebrate animals, play one of the most important parts in the "non-miraculous history of creation." For these remains and impressions of extinct animals and plants, preserved to us in a petrified condition, are the true "monuments of the creation," the infallible and indisputable records which fix the correct history of organisms upon an irrefragable foundation. All petrified or fossil remains and impressions tell us of the forms and structure of such animals and plants as are either the progenitors and ancestors of the present living organisms, or they are the representatives of extinct collateral lines, which, together with the present living organisms, branched off from a common stem.

These inestimable records of the history of creation throughout a long period played a subordinate part in science. Their true nature was indeed correctly understood, even more than five hundred years before Christ, by the great Greek philosopher, Xenophanes of Colophon, the same who founded the so-called Eleatic philosophy, and who was the first to demonstrate with convincing precision that all conceptions of personal gods result in more or less rude anthropomorphism.

Xenophanes for the first time, asserted that the fossil impressions of animals and plants were real remains of formerly living creatures, and that the mountains in whose rocks they were found must at an earlier date have stood under water. But although other great philosophers of antiquity, and among them Aristotle, also possessed this true knowledge, yet throughout the illiterate Middle Ages, and even with some naturalists of the last century, the idea prevailed that petrifactions were so-called freaks of nature (lusus naturae), or products of an unknown formative power or instinct of nature (nisus formativus, vis plastica). Respecting the nature of this mysterious and mystic creative power, the strangest ideas were formed. Some believed that this constructive power—the same to which they also ascribed the coming into existence of the present species of animals and plants—had made numerous attempts to create organisms of different forms, but that these attempts had only partially succeeded, had often failed, and that petrifactions were nothing more than such unsuccessful attempts. According to others, petrifactions originated from the influence of the stars upon the interior of the earth.

Others, again, had the still cruder notion that the Creator had first made models (out of mineral substances—for example, of gypsum or clay) of those forms of animals and plants which he afterwards executed in organic substances, and into which he breathed his living breath; petrifactions were accordingly such rude inorganic models. Even as late as the last century these crude ideas prevailed, and it was assumed, for example, that there existed a special "seminal air," which was said to penetrate into the earth with the water, and by fructifying the stones formed petrifactions or "stony flesh" (caro fossilis).

It took a very long time before the simple and natural view was accepted, namely, that petrifactions are in reality nothing but what they appear to simple observation—the indestructible remains of extinct organisms. It is true the celebrated painter, Leonardo da Vinci, in the 15th century, ventured to assert that the mud which was constantly deposited by water was the cause of petrifactions, as it surrounded the indestructible shells of mussels and snails which lay at the bottom of the waters, and gradually turned them into solid stone. The same idea was maintained in the 16th century by a Parisian potter, Palissy by name, who became celebrated on account of his invention of china. However, the so-called "professional men" were very far from paying any regard to these correct assertions of a simple and healthy human understanding; it was not till the end of the last century that it was generally accepted, in consequence of the foundation of the Neptunian geology by Werner.

The foundation of a more strictly scientific palaeontology, however, belongs to the beginning of our century, when Cuvier published his classic researches on petrified Vertebrate animals, and when his great opponent, Lamarck, made known his remarkable investigations on fossil Invertebrate animals, especially on petrified snails and clams. In Cuvier's celebrated work "On the Fossil Bones" of Vertebrate animals—principally of mammals and reptiles—we see that he had already arrived at the knowledge of some very important and general palaeontological laws, which are of great consequence to the history of creation. Foremost among them stands the assertion that the extinct species of animals, whose remains we find petrified in the different strata of the earth's crust, lying one above another, differ all the more strikingly from the still living kindred species of animals the deeper those strata lie—in other words, the earlier the animals lived in past ages. In fact, in every perpendicular section of the stratified crust of the earth we find that the different strata, deposited by the water in a certain historical succession, are characterized by different petrifactions, and that these extinct organisms become more like those of the present day the higher the strata lie; in other words, the more recent the period in the earth's history in which they lived, died, and became encrusted by the deposited and hardened strata of mud.

However important this general observation of Cuvier's was in one sense, yet in another it became to him the source of a very serious error. For as he considered the characteristic petrifactions of each individual group of strata (which had been deposited during one main period of the earth's history) to be entirely different from those of the strata lying above or below, and as he erroneously believed that one and the same species of animal was never found in two succeeding groups of strata, he arrived at the false idea, which was accepted as a law by most subsequent naturalists, that a series of quite distinct periods of creation had succeeded one another. Each period was supposed to have had its special animal and vegetable world, each its peculiar specific Fauna and Flora.

Cuvier imagined that the whole history of the earth's crust, since the time when living creatures had first appeared on the surface, must be divided into a number of perfectly distinct periods, or divisions of time, and that the individual periods must have been separated from one another by peculiar revolutions of an unknown nature (cataclysms, or catastrophes). Each revolution was followed by the utter annihilation of the till then existing animals and plants, and after its termination a completely new creation of organic forms took place. A new world of animals and plants, absolutely and specifically distinct from those of the preceding historical periods, was called into existence at once, and now again peopled the globe for thousands of years, till it again perished suddenly in the crash of a new revolution.

About the nature and causes of these revolutions, Cuvier expressly said that no idea could be formed, and that the present active forces in nature were not sufficient for their explanation. Cuvier points out four active causes as the natural forces, or mechanical agents, at present constantly but slowly at work in changing the earth's surface: first, *rain*, which washes down the steep mountain slopes and heaps up debris at their foot; secondly, *flowing waters*, which carry away this debris and deposit it as mud in stagnant waters; thirdly, the sea, whose breakers gnaw at the steep *sea* coasts, and throw up "dunes" on the flat sea margins; finally and fourthly, *volcanos*, which break through and heave up the strata of the earth's hardened crust, and pile up and scatter about the products of their eruptions. Whilst Cuvier recognizes the constant slow transformation of the present surface of the earth by these four mighty causes, he asserts at the same time that they would not have sufficed to effect the revolutions of the remote ages, and that the anatomical structure of the earth's surface cannot be explained by the necessary action of those mechanical agents: the great and marvellous revolutions of the whole earth's surface must, according to him, have been rather the effects of very peculiar causes, completely unknown to us; the usual thread of development was broken by them, and the course of nature altered.

These views Cuvier explained in a special work "On the Revolutions of the Earth's Surface, and the Changes which they have wrought in the Animal World." They were maintained, and generally accepted for a long time, and became the greatest obstacle to the development of a natural history of the creation. For if such all-destructive revolutions had actually occurred, of course a continuity of the development of species, a connecting thread in the organic history of the earth, could not be admitted at all, and we should be obliged to have recourse to the action of supernatural forces; that is, to the interference of miracles in the natural course of things. It is only through miracles that these revolutions of the earth could have been brought about, and it is only through miracles that, after their cessation and at the commencement of each new period, a new animal and vegetable kingdom could have been created. But science has no room for miracles, for by miracles we understand an interference of supernatural forces in the natural course of development of matter.

Just as the great authority which Linnaeus gained by his system of distinguishing and naming organic species led his successors to a complete ossification, as it were, of the dogmatic idea of species and to a real abuse of the systematic distinction implied by it, so the great services which Cuvier had rendered to the knowledge and distinction of extinct species became the cause of a general adoption of his theory of revolutions and catastrophes, and of the false views of creation connected therewith. The consequence of this was that, during the first half of our century, most zoologists and botanists clung to the opinion that a series of independent periods in the organic history of the earth had existed; that each period was distinguished by distinct and peculiar kinds of animal and vegetable species; that these were annihilated at the termination of the period by a general revolution; and that, after the cessation of the latter, a new world of different species of animals and plants was created.

It is true some independent thinkers, above all the great physical philosopher, Lamarck, even at an early period, set forth a series of weighty reasons which refuted Cuvier's theory of cataclysms, and pointed to a perfectly continuous and uninterrupted developmental history of all the organic inhabitants of the earth through all ages. They maintained that the animal and vegetable species of each period were derived from those of the preceding period, and were only the altered descendants of the former. This true conception, however, being opposed to Cuvier's great authority, was then unable to make way. Nay, even after Cuvier's theory of catastrophes had been completely cast out from the domain of geology by Lyell's classic Principles of Geology, which appeared in 1830, still his idea of the specific distinctness of a series of organic creations maintained its influence, in many ways, in the science of Palaeontology. (Gen. Morph. ii. 312.)

By a curious coincidence, thirteen years ago, almost at the same time that Cuvier's History of Creation received its death-blow by Darwin's book, another celebrated naturalist made an attempt to re-establish it, and to adopt it in the roughest manner, as a part of a teleologico-theological system of nature. This was the Swiss geologist, Louis Agassiz, who attained a great reputation by his theory of glaciers and the ice-period, borrowed from Schimper and Charpentier, and who has been living in North America for many years. He commenced in 1858 to publish a work planned on a very large scale, which bears the title of "Contributions to the Natural History of the United States of North America." The first volume of this work, although large and costly, owing to the patriotism of the Americans, had an unprecedented sale; its title is, "An Essay on Classification."[5]

In this essay Agassiz not only discusses the natural series of organisms, and the different attempts of naturalists at classification, but also all the general biological phenomena which have reference to it. The history of the development of organisms, both the embryonal and the palaeontological, comparative anatomy, the general economy of nature, the geographical and topographical distribution of animals and plants—in short, almost all the general phenomena of organic nature are discussed in Agassiz's Essay on Classification, and are explained in a sense and from a point of view which is thoroughly opposed to that of Darwin. While Darwin's chief merit lies in the fact that he demonstrates natural causes for the coming into existence of animal and vegetable species, and thereby establishes the mechanical or monistic view of the universe as regards this most difficult branch of the history of creation, Agassiz, on the contrary, strives to exclude every mechanical hypothesis from the subject, and to put the supernatural interference of a personal Creator in the place of the natural forces of matter; consequently, to establish a thoroughly teleological or dualistic view of the universe. It will not be out of place if I examine a little more closely Agassiz's biological views, and especially his ideas of creation, because no other work of our opponents treats the important fundamental questions with equal minuteness, and because the utter untenableness of the dualistic conception of nature becomes very evident from the failure of this attempt.

The organic *species*, the various conceptions of which we have above designated as the real centre of dispute in the opposed views of creation, is looked upon by Agassiz, as by Cuvier and Linnaeus, as a form unchangeable in all its essential characteristics. The species may indeed change and vary within certain narrow limits; never in essential qualities, but only in unessential points. No new species could ever proceed from the changes or varieties of a species. Not one of all organic species, therefore, is ever derived from another, but each individual species has been separately created by God. Each individual species, as Agassiz expresses it, is "an embodied creative thought" of God.

In direct opposition to the fact established by palaeontological experience, that the duration of the individual organic species is most unequal, and that many species continue unchanged through several successive periods of the earth's history, while others only existed during a small portion of such a period, Agassiz maintains that one and the same species never occurs in two different periods, but that each individual period is characterized by species of animals and plants which are quite peculiar, and belong to it exclusively. He further shares Cuvier's opinion that the whole of these inhabitants were annihilated by the great and universal revolutions of the earth's surface, which divide two successive periods, and that after its destruction a new and specifically different assemblage of organisms was created. This new creation Agassiz supposes to have taken place in this manner: viz., that at each creation all the inhabitants of the earth, in their full average number of individuals, and in the peculiar relations corresponding to the economy of nature, were, as a whole, suddenly placed upon the earth by the Creator. In saying this he puts himself in opposition to one of the most firmly established and most important laws of animal and vegetable geography—namely, to the law that each species has a single original locality of origin, or a so-called "centre of creation," from which it has gradually spread over the rest of the earth. Instead of this, Agassiz assumes each species to have been created at several points of the earth's surface, and that in each case a large number of individuals was created.

The "natural system" of organisms, the different groups and categories of which arranged above one another—namely, the branches, classes, orders, families, genera, and species—we consider, in accordance with the Theory of Descent, as different branches and twigs of the organic family-tree, is, according to Agassiz, the direct expression of the divine plan of creation, and the naturalist, while investigating the natural system, repeats the creative thoughts of God. In this Agassiz finds the strongest proof that man is the image and child of God. The different stages of groups or categories of the natural system correspond with the different stages of development which the divine plan of creation had attained. The Creator, in projecting and carrying out this plan, starting from the most general ideas of creation, plunged more and more into specialities. For instance, when creating the animal kingdom, God had in the first place four totally distinct ideas of animal bodies, which he embodied in the different structures of the four great, principal forms, types, or branches of the animal kingdom; namely, vertebrate animals, articulate animals, molluscous animals, and radiate animals. The Creator then, having reflected in what manner he might vary these four different plans of structure, next created within each of the four principal forms, several different classes—for example, in the vertebrate animal form, the classes of mammals, birds, reptiles, amphibious animals, and fishes. Then God further reflected upon the individual classes, and by various modifications in the structure of each class, he produced the individual orders. By further variation in the order, he created natural families. As the Creator further varied the peculiarities of structure of individual parts in each family, genera arose. In further meditation on his plan of creation, he entered so much into detail that individual species came into existence, which, consequently, are embodied creative thoughts of the most special kind. It is only to be regretted that the Creator expressed these most special and most deeply considered "creative thoughts" in so very indistinct and loose a manner, and that he imprinted so vague a stamp upon them, and permitted them to vary so freely that not one naturalist is able to distinguish the "good" from the "bad species," or a genuine species from varieties, races, etc. (Gen. Morph. ii. 373.)

We see, then, according to Agassiz's conception, that the Creator, in producing organic forms, goes to work exactly as a human architect, who has taken upon himself the task of devising and producing as many different buildings as possible, for the most manifold purposes, in the most different styles, in various degrees of simplicity, splendour, greatness, and perfection. This architect would perhaps at first choose four different styles for all these buildings, say the Gothic, Byzantine, Chinese, and Rococo styles. In each of these styles he would build a number of churches, palaces, garrisons, prisons, and dwelling-houses. Each of these different buildings he would execute in ruder and more perfect, in greater and smaller, in simpler and grander fashion, etc. However, the human architect would perhaps, in this respect, be better off than the divine Creator, as he would have perfect liberty in the number of graduated subordinate groups. The Creator, however, according to Agassiz, can only move within six groups or categories: the species, genus, family, order, class, and type. More than these six categories do not exist for him.

When we read Agassiz's book on classification, and see how he carries out and establishes these strange ideas, we can scarcely understand how, with all the appearance of scientific earnestness, he can persevere in his idea of the divine Creator as a man-like being (anthropomorphism), for by his explanation of details he produces a picture of the most absurd nonsense. In the whole series of these suppositions the Creator is nothing but an almighty man, who, plagued with *ennui*, amuses himself with planning and constructing most varied toys in the shape of organic species. After having diverted himself with these for thousands of years, they become tiresome to him, he destroys them by a general revolution of the earth's surface, and thus throws the whole of the useless toys in heaps together; then, in order to while away his time with something new and better, he calls a new and more perfect animal and vegetable world into existence. But in order not to have the trouble of beginning the work of creation over again, he keeps, in the main, to his original plan of creation, and creates merely new species, or at most only new genera, and much more rarely new families, new orders, or classes. He never succeeds in producing a new style or type, and always keeps strictly within the six categories or graduated groups.

When, according to Agassiz, the Creator has thus amused himself for thousands of millions of years with constructing and destroying a series of different creations, at last (but very late) he is struck with the happy thought of creating something like himself, and so makes man in his own image. The end of all the history of creation is thus arrived at and the series of revolutions of the earth is closed. Man, the child and image of God, gives him so much to do, causes him so much pleasure and trouble, that he is wearied no longer, and therefore need not undertake a new creation. It is clear that if, according to Agassiz, we once assign to the Creator entirely human attributes and qualities, and regard his work of creation as entirely analogous to human creative activity, we are necessarily obliged to admit such utterly absurd inferences as those just stated.

The many intrinsic contradictions and perversities in Agassiz's view of creation—a view which necessarily led him to the most decided opposition to the Theory of Descent—must excite our astonishment all the more because, in his earlier scientific works, he had in many respects actually paved the way for Darwin, especially by his researches in Palaeontology. Among the numerous investigations which created general interest in the then young science of Palaeontology, those of Agassiz, especially his celebrated work on "Fossil Fish," rank next in importance to Cuvier's work, which formed the foundation of the science. The petrified fish, with which Agassiz has made us acquainted, have not only an extremely great importance for the understanding of all groups of Vertebrate animals, and their historical development, but we have arrived through them at a sure knowledge of important general laws of development, some of which were first discovered by Agassiz. He it was who drew special attention to the remarkable parallelism between the embryonal and the palaeontological development—between ontogeny and phylogeny, which I have already (p. 10 [5]) claimed as one of the strongest pillars of the Theory of Descent. No one before had so distinctly stated as Agassiz did, that, of the Vertebrate animals, fishes alone existed, at first, that amphibious animals came later, and that birds and mammals appeared only at a much later period, further, that among mammals, as among fishes, imperfect and lower orders had appeared first, but more perfect and higher orders at a later period. Agassiz, therefore, showed that the palaeontological development of the whole Vertebrate group was not only parallel with the embryonic, but also with the systematic development, that is, with the graduated series which we see everywhere in the system, ascending from the lower to the higher classes, orders, etc.

In the earth's history lower forms appeared first, the higher forms later. This important fact, as well as the agreement of the embryonic and palaeontological development, is explained quite simply and naturally by the Doctrine of Descent, and without it is perfectly inexplicable. This cause holds good also in the great law of *progressive development*, that is, of the historical progress of organization, which is traceable, broadly and as a whole, in the historical succession of all organisms, as well as in the special perfecting of individual parts of animal bodies. Thus, for example, the skeleton of Vertebrate animals acquired at first slowly, and by degrees, that high degree of perfection which it now possesses in man and the other higher Vertebrate animals. This progress, acknowledged in point of fact by Agassiz, necessarily follows from Darwin's Doctrine of Descent, which demonstrates its active causes. If this doctrine is correct, the perfecting and diversification of animal and vegetable species must of necessity have gradually increased in the course of the organic history of the earth, and could only attain its highest perfection in most recent times.

The above-mentioned laws of development, together with some other general ones, which have been expressly admitted and justly emphasized by Agassiz, and some of which have first been set forth by him, are, as we shall see later, only explicable by the Theory of Descent, and without it remain perfectly incomprehensible. The conjoint action of Inheritance and Adaptation, as explained by Darwin, can alone be their true cause. But they all stand in sharp and irreconcilable opposition to the hypothesis of creation maintained by Agassiz, as well as to the idea of a personal Creator who acts for a definite purpose. If we seriously wish to explain those remarkable phenomena and their inter-connection by Agassiz's theory, then we are necessarily driven to the curious supposition that the Creator himself has developed, together with the organic nature which he created and modelled. We can, in that case, no longer rid ourselves of the idea that the Creator himself, like a human being, designed, improved, and finally, with many alterations, carried out his plans. "Man grows as higher grow his aims," and the same supposition, so unworthy of a God, must be applied to him. Although, from the reverence with which, in every page, Agassiz speaks of the Creator, it might appear that, on his theory, we attain to the sublimest conception of the divine activity in nature, yet the contrary is in truth the case. The divine Creator is degraded to the level of an idealized man, of an organism progressing in development!

Considering the wide popularity and great authority which Agassiz's work has gained, and which is perhaps justified on account of earlier scientific services rendered by the author, I have thought it my duty here to show the utter untenableness of his general conceptions. So far as this work pretends to be a scientific history of creation, it is undoubtedly a complete failure. But still it has great value, being the only detailed attempt, adorned with scientific arguments, which an eminent naturalist of our day has made to found a teleological or dualistic history of creation. The utter impossibility of such a history has thus been made obvious to every one. No opponent of Agassiz could have refuted the dualistic conception of organic nature and its origin more strikingly than he himself has done by the intrinsic contradictions which present themselves everywhere in his theory.

The opponents of the monistic or mechanical conception of the world have welcomed Agassiz's work with delight, and find in it a perfect proof of the direct creative action of a personal God. But they overlook the fact that this personal Creator is only an idealized organism, endowed with human attributes. This low dualistic conception of God corresponds with a low animal stage of development of the human organism. The more developed man of the present day is capable of, and justified in, conceiving that infinitely nobler and sublimer idea of God which alone is compatible with the monistic conception of the universe, and which recognizes God's spirit and power in all phenomena without exception. This monistic idea of God, which belongs to the future, has already been expressed by Giordano Bruno in the following words:—"A spirit exists in all things, and no body is so small but contains a part of the divine substance within itself, by which it is animated." It is of this noble idea of God that Goethe says:—"Certainly there does not exist a more beautiful worship of God than that which needs no image, but which arises in our heart from converse with Nature." By it we arrive at the sublime idea of the Unity of God and Nature.

CHAPTER IV.

THEORY OF DEVELOPMENT ACCORDING TO GOETHE AND OKEN.

Scientific Insufficiency of all Conceptions of a Creation of Individual Species.—Necessity of the Counter Theories of Development.—Historical Survey of the Most Important Theories of Development.—Aristotle.—His Doctrine of Spontaneous Generation.—The Meaning of Natural Philosophy.—Goethe.—His Merits as a Naturalist.—His Metamorphosis of Plants.—His Vertebral Theory of the Skull.—His Discovery of the Mid Jawbone in Man.—Goethe's Interest in the Dispute between Cuvier and Geoffroy St. Hilaire.—Goethe's Discovery of the Two Organic Formative Principles, of the Conservative Principle of Specification (by Inheritance), and of the Progressive Principle of Transformation (by Adaptation).—Goethe's Views of the Common Descent of all Vertebrate Animals, including Man.—Theory of Development according to Gottfried Reinhold Treviranus.—His Monistic Conception of Nature.—Oken.—His Natural Philosophy.—Oken's Theory of Protoplasm.—Oken's Theory of Infusoria (Cell Theory).—Oken's Theory of Development.

ALL the different ideas which we may form of a separate and independent origin of the individual organic species by creation lead us, when logically carried out, to a so-called *anthropomorphism*, that is, to imagining the Creator as a man-like being, as was shown in our last chapter. The Creator becomes an organism who designs a plan, reflects upon and varies this plan, and finally forms creatures according to this plan, as a human architect would his building. If even such eminent naturalists as Linnaeus, Cuvier, and Agassiz, the principal representatives of the dualistic hypothesis of creation, could not arrive at a more satisfactory view, we may take it as evidence of the insufficiency of all those conceptions which would derive the various forms of organic nature from a creation of individual species.

Some naturalists, indeed, seeing the complete insufficiency of these views, have tried to replace the idea of a personal Creator by that of an unconsciously active and creative Force of Nature; yet this expression is evidently merely an evasive phrase, as long as it is not clearly shown what this force of nature is, and how it works. Hence these attempts, also, have been absolute failures. In fact, whenever an independent origin of the different forms of animals and plants has been assumed, naturalists have found themselves compelled to fall back upon so many "acts of creation," that is, on supernatural interferences of the Creator in the natural course of things, which in all other cases goes on without interference.

It is true that several teleological naturalists, feeling the scientific insufficiency of a supernatural *"creation,"* have endeavoured to save the hypothesis by wishing it to be understood that creation "is nothing else than a way of coming into being, unknown and inconceivable to us." The eminent Fritz Müller has cut off from this sophistic evasion every chance of escape by the following striking remark:—"It is intended here only to express in a disguised manner the shamefaced confession, that they neither have, nor care to have, *any opinion* about the origin of species. According to this explanation of the word, we might as well speak of the creation of cholera, or syphilis, of the creation of a conflagration, or of a railway accident, as of the creation of man." (Jenaische Zestscrift, bd. v. p. 272.)

In the face, then, of these hypotheses of creation, which are scientifically insufficient, we are forced to seek refuge in the *counter-theory of development* of organisms, if we wish to come to a rational conception of the origin of organisms. We are forced and obliged to do so, even if the theory of development only throws a glimmer of probability upon a mechanical, natural origin of the animal and vegetable species; but all the more if, as we shall see, this theory explains all facts simply and clearly, as well as completely and comprehensively. The theories of development are by no means, as they often falsely are represented to be, arbitrary fancies, or wilful products of the imagination, which only attempt approximately to explain the origin of this or that individual organism; but they are theories founded strictly on science, which explain in the simplest manner, from a fixed and clear point of view, the whole of organic natural phenomena, and more especially the origin of organic species, and demonstrate them to be the necessary consequences of mechanical processes in nature.

As I have already shown in the second chapter, all these theories of development coincide naturally with that general theory of the universe which is usually designated as the uniform or *monistic*, often also as the *mechanical* or causal, because it only assumes mechanical causes, or *causes working by necessity* (causae efficientes), for the explanation of natural phenomena. In like manner, on the other hand, the supernatural hypotheses of creation which we have already discussed coincide completely with the opposite view of the universe, which in contrast to the former is called the twofold or *dualistic*, often the *teleological* or vital, because it traces the organic natural phenomena to final causes, acting and *working for a definite purpose* (causae finales). It is this deep and intrinsic connection of the different theories of creation with the most important questions of philosophy that incites us to their closer examination.

The fundamental idea, which must necessarily lie at the bottom of all natural theories of development, is that of a *gradual development of all (even the most perfect) organisms* out of a single, or out of a very few, quite simple, and quite imperfect original beings, which came into existence, not by supernatural creation, but by *spontaneous generation*, or archigony, out of inorganic matter. In reality, there are two distinct conceptions united in this fundamental idea, but which have, nevertheless, a deep intrinsic connection—namely, first, the idea of spontaneous generation (or archigony) of the original primary beings; and secondly, the idea of the progressive development of the various species of organisms from those most simple primary beings. These two important mechanical conceptions are the inseparable fundamental ideas of every theory of development, if scientifically carried out. As it maintains the derivation of the different species of animals and plants from the simplest, common primary species, we may term it also the Doctrine of Filiation, or *Theory of Descent*; as there is also a change of species connected with it, it may also be termed the *Transmutation Theory*.

While the supernatural histories of creation must have originated thousands of years ago, in that very remote primitive age when man, first developing out of the monkey-state, began for the first time to think more closely about himself, and about the origin of the world around him, the natural theories of development, on the other hand, are necessarily of much more recent origin. These views are met with only among nations of a more matured civilization, to whom, by philosophic culture, the necessity of a knowledge of natural causes has become apparent; and even among these, only individual and specially gifted natures can be expected to have recognized the origin of the world of phenomena, as well as its course of development, as the necessary consequences of mechanical, naturally active causes. In no nation have these preliminary conditions, for the origin of a natural theory of development, ever existed in so high a degree as among the Greeks of classic antiquity. But, on the other hand, they lacked a close acquaintance with the facts of the processes and forms of nature, and, consequently, the foundation based upon experience, for a satisfactory unravelling of the problem of development. Exact investigation of nature, and the knowledge of nature founded on an experimental basis, was of course almost unknown to antiquity, as well as to the Middle Ages, and is only an acquisition of modern times. We have therefore here no special occasion to examine the natural theories of development of the various Greek philosophers, since they were wanting in the knowledge gained by experience, both of organic and inorganic nature, and since they almost always, as the consequence, lost themselves in airy speculations.

One man only must be mentioned here by way of exception,—Aristotle, the greatest and the only truly great naturalist of antiquity and the Middle Ages, one of the grandest geniuses of all time. To what a degree he stands there alone, during a period of more than two thousand years, in the region of empirico-philosophical knowledge of nature, and especially in his knowledge of organic nature, is proved to us by the precious remains of his but partially surviving works. In them many traces are found of a theory of natural development. Aristotle assumes, as a matter of certainty, that spontaneous generation was the natural manner in which the lower organic creatures came into existence. He describes animals and plants originating from matter itself, through its own original force; as, for example, moths from wool, fleas from putrid dung, wood-lice from damp wood, etc. But as the distinction of organic species, which Linnaeus only arrived at two thousand years later, was unknown to him, he could form no ideas about their genealogical relations.

The fundamental notion of the theory of development, that the different species of animals and plants have been developed from a common primary species by transformation, could of course only be clearly asserted after the kinds of species themselves had become better known, and after the extinct species had been carefully examined and compared with the living ones. This was not done until the end of the last and the beginning of the present century. It was not until the year 1801 that the great Lamarck expressed the theory of development, which he, in 1809, further elaborated in his classical "Philosophie Zoologique." While Lamarck and his countryman, Geoffroy St. Hilaire, in France, opposed Cuvier's views, and maintained a natural development of organic species by transformation and descent, Goethe and Oken at the same time pursued the same course in Germany, and helped to establish the theory of development. As these naturalists are generally called nature-philosophers (Naturphilosophen), and as this ambiguous designation is correct in a certain sense, it appears to me appropriate here to say a few words about the correct estimate of the "Natur-philosophie."

Although for many years in England the ideas of natural science and philosophy have been looked upon as almost equivalent, and as every truly scientific investigator of nature is most justly called there a "natural philosopher," yet in Germany for more than half a century natural science has been kept strictly distinct from philosophy, and the union of the two into a true philosophy of nature is recognized only by the few. This misapprehension is owing to the fantastic eccentricities of earlier German natural-philosophers, such as Oken, Schelling, etc.; they believed that they were able to construct the laws of nature in their own heads, without being obliged to take their stand upon the grounds of actual experience. When the complete hollowness of their assumptions had been demonstrated, naturalists, in "the nation of thinkers," fell into the very opposite extreme, believing that they would be able to reach the high aim of science, that is, the knowledge of truth, by the mere experience of the senses, and without any philosophical activity of thought.

From that time, but especially since 1830, most naturalists have shown a strong aversion to any general, philosophical view of nature. The real aim of natural science was now supposed to consist in the knowledge of details, and it was believed that this would be attained in the study of biology, when the forms and the phenomena of life, in all individual organisms, had become accurately known, by the help of the finest instruments and means of observation. It is true that among these strictly empirical, or so-called exact naturalists, there were always very many who rose above this narrow point of view, and sought the final aim in a knowledge of the general laws of organization. Yet the great majority of zoologists and botanists, during the thirty or forty years preceding Darwin, refused to concern themselves about such general laws; all they admitted was, that perhaps in the far distant future, when the end of all empiric knowledge should have been arrived at, when all individual animals and plants should have been thoroughly examined, naturalists might begin to think of discovering general biological laws.

If we consider and compare the most important advances which the human mind has made in the knowledge of truth, we shall soon see that it is always owing to philosophical mental operations that these advances have been made, and that the experience of the senses which certainly and necessarily precedes these operations, and the knowledge of details gained thereby, only furnish the basis for those general laws. Experience and philosophy, therefore, by no means stand in such exclusive opposition to each other as most men have hitherto supposed; they rather necessarily supplement each other. The philosopher who is wanting in the firm foundation of sensuous experience, of empirical knowledge, is very apt to arrive at false conclusions in his general speculations, which even a moderately informed naturalist can refute at once. On the other hand, the purely empiric naturalists, who do not trouble themselves about the philosophical comprehension of their sensuous experiences, and who do not strive after general knowledge, can promote science only in a very slight degree, and the chief value of their hard-won knowledge of details lies in the general results which more comprehensive minds will one day derive from them.

From a general survey of the course of biological development since Linnaeus' time, we can easily see, as Bär has pointed out, a continual vacillation between these two tendencies, at one time a prevalence of the empirical—the so-called exact—and then again of the philosophical or speculative tendency. Thus at the end of the last century, in opposition to Linnaeus' purely empirical school, a natural-philosophical reaction took place, the moving spirits of which, Lamarck, Geoffroy St. Hilaire, Goethe, and Oken, endeavoured by their mental work to introduce light and order into the chaos of the accumulated empirical raw material. In opposition to the many errors and speculations of these natural philosophers, who went too far, Cuvier then came forward, introducing a second, purely empirical period. It reached its most one-sided development between the years 1830–1860, and there now followed a second philosophical reaction, caused by Darwin's work. Thus during the last ten years, men again have begun to endeavour to obtain a knowledge of the general laws of nature, to which, after all, all detailed knowledge of experience serves only as a foundation, and through which alone it acquires its true value. It is through philosophy alone that natural knowledge becomes a true science, that is, a philosophy of nature. (Gen. Morph. i. 63–108.)

Jean Lamarck and Wolfgang Goethe stand at the head of all the great philosophers of nature who first established a theory of organic development, and who are the illustrious fellow-workers of Darwin. I turn first to our beloved Goethe, who, among all, stands in the closest relations to us Germans. However, before I explain his special services to the theory of development, it seems to me necessary to say a few words about his importance as a naturalist in general, as it is commonly very little known.

I am sure most of my readers honour Goethe only as a poet and a man; only a few have any conception of the high value of his scientific works, and of the gigantic stride with which he advanced before his own age—advanced so much that most naturalists of that time were unable to follow him. In several passages of his scientific writings he bitterly complains of the narrow-mindedness of professed naturalists, who do not know how to value his works (who cannot see the wood for the trees), and who cannot rouse themselves to discover the general laws of nature among the mass of details. He is only too just when he utters the reproach—"The philosophers will very soon discover that observers rarely rise to a stand-point from which they can survey so many important objects." It is true, at the same time, that their want of appreciation was caused by the false road into which Goethe was led in his theory of colours.

This theory of colours, which he himself designates as the favourite production of his leisure, however much that is beautiful it may contain, is a complete failure in regard to its foundations. The exact mathematical method by means of which alone it is possible, in inorganic sciences, but above all in physics, to raise a structure step by step on a thoroughly firm basis, was altogether repugnant to Goethe. In rejecting it he allowed himself not only to be very unjust towards the most eminent physicists, but to be led into errors which have greatly injured the fame of his other valuable works. It is quite different in the organic sciences, in which we are but rarely able to proceed, from the beginning, upon a firm mathematical basis; we are rather compelled, by the infinitely difficult and intricate nature of the problem, at the first to form inductions—that is, we are obliged to endeavour to establish general laws by numerous individual observations, which are not quite complete. A comparison of kindred series of phenomena, or the method of combination, is here the most important instrument for inquiry, and this method was applied by Goethe with as much success as with conscious knowledge of its value, in his works relating to the philosophy of nature.

The most celebrated among Goethe's writings relating to organic nature is his *Metamorphosis of Plants*, which appeared in 1790, a work which distinctly shows a grasp of the fundamental idea of the theory of development, inasmuch as Goethe, in it, was labouring to point out a single organ, by the infinitely varied development and metamorphosis of which the whole of the endless variety of forms in the world of plants might be conceived to have arisen; this fundamental organ he found in the *leaf.* If at that time the microscope had been generally employed, if Goethe had examined the structure of organisms by the means of the microscope he would have gone still further, and would have seen that the leaf is itself a compound of individual parts of a lower order, that is, of *cells.* He would then not have declared that the leaf, but that the *cell* is the real fundamental organ by the multiplication, transformation, and combination (synthesis) of which, in the first place, the leaf is formed; and that, in the next place, by transformation, variation, and combination of leaves there arise all the varied beauties in form and colour which we admire in the green parts, as well as in the organs of propagation, or the flowers of plants. Goethe here showed that in order to comprehend the whole of the phenomena, we must in the first place compare them, and, secondly, search for a simple type, a simple fundamental form, of which all other forms are only infinite variations.

Something similar to what he had here done for the metamorphosis of plants he then did for the Vertebrate animals, in his celebrated *vertebral theory of the skull.* Goethe was the first to show, independently of Oken, who almost simultaneously arrived at the same thought, that the skull of man and of all Vertebrate animals, in particular mammals, is nothing more than a bony case, formed of the same bones,—that is, vertebrae,—out of which the spine also is composed. The vertebrae of the skull are like those of the spine, bony rings lying behind each other, but in the skull are peculiarly changed and specialized (differentiated). Although this idea has been strongly modified by recent discoveries, yet in Goethe's day it was one of the greatest advances in comparative anatomy, and was not only one of the first advances towards the understanding of the structure of Vertebrate animals, but at the same time explained many individual phenomena. When two parts of a body, such as the skull and spine, which appear at first sight so different, were proved to be parts originally the same, developed out of one and the same foundation, one of the difficult problems of the philosophy of nature was solved. Here again we meet the notion of a single type—the conception of a single principle, which becomes infinitely varied in the different species, and in the parts of individual species.

But Goethe did not merely endeavour to search for such far-reaching laws, he also occupied himself most actively for a long time with numerous individual researches, especially in comparative anatomy. Among these, none is perhaps more interesting than the discovery of the *mid jawbone in man.* As this is, in several respects, of importance to the theory of development, I shall briefly explain it here. There exist in all mammals two little bones in the upper jaw, which meet in the centre of the face, below the nose, and which lie between the two halves of the real upper jawbone. These two bones, which hold the four upper cutting teeth, are recognized without difficulty in most mammals; in man, however, they were at that time unknown, and celebrated comparative anatomists even laid great stress upon this want of a mid jawbone, as they considered it to constitute the principal difference between men and apes—the want of a mid jawbone was, curiously enough, looked upon as the most human of all human characteristics. But Goethe could not accept the notion that man, who in all other corporeal respects was clearly only a mammal of higher development, should lack this mid jawbone.

By the general law of induction as to the mid jawbone he arrived at the special deductive conclusion that it must exist in man also, and Goethe did not rest until, after comparing a great number of human skulls, he really found the mid jawbone. In some individuals it is preserved throughout a whole lifetime, but usually at an early age it coalesces with the neighbouring upper jawbone, and is therefore only to be found as an independent bone in very youthful skulls. In human embryos it can now be pointed out at any moment. In man, therefore, the mid jawbone actually exists, and to Goethe the honour is due of having first firmly established this fact, so important in many respects; and this he did while opposed by the celebrated anatomist, Peter Camper, one of the most important professional authorities. The way by which Goethe succeeded in establishing this fact is especially interesting; it is the way by which we continually advance in biological science, namely, by way of induction and deduction. *Induction* is the inference of a general law from the observation of numerous individual cases; *deduction*, on the other hand, is an inference from this general law applied to a single case which has not yet been actually observed. From the collected empirical knowledge of those days, the inductive conclusion was arrived at that all mammals had mid jawbones. Goethe drew from this the deductive conclusion, that man, whose organization was in all other respects not essentially different from mammals, must also possess this mid jawbone; and on close examination it was actually found. The deductive conclusion was confirmed and verified by experience.

Even these few remarks may serve to show the great value which we must ascribe to Goethe's biological researches. Unfortunately most of his labours devoted to this subject are so hidden in his collected works, and his most important observations and remarks so scattered in numerous individual treatises—devoted to other subjects—that it is difficult to find them out. It also sometimes happens that an excellent, truly scientific remark is so much interwoven with a mass of useless philosophical fancies, that the latter greatly detract from the former.

Nothing is perhaps more characteristic of the extraordinary interest which Goethe took in the investigation of organic nature than the lively way in which, even in his last years, he followed the dispute which broke out in France between Cuvier and Geoffroy de St. Hilaire. Goethe, in a special treatise which was only finished a few days before his death, in March,

1832, has given an interesting description of this remarkable dispute and its general importance, as well as an excellent sketch of the two great opponents. This treatise bears the title "Principes de Philosophic Zoologique par M. Geoffroy de Saint Hilaire"; it is Goethe's last work, and forms the conclusion of the collected edition of his works. The dispute itself was, in several respects, of the highest interest. It turned essentially upon the justification of the theory of development. It was carried on, moreover, in the bosom of the French Academy, by both opponents, with a personal vehemence almost unheard of in the dignified sessions of that learned body; this proved that both naturalists were fighting for their most sacred and deepest convictions. The conflict began on the 22nd of February, and was followed by several others; the fiercest took place on the 19th of July, 1830. Geoffroy, as the chief of the French nature-philosophers, represented the theory of natural development and the monistic conception of nature. He maintained the mutability of organic species, the common descent of the individual species from common primary forms, and the unity of their organization—or the unity of the plan of structure, as it was then called.

Cuvier was the most decided opponent of these views, and according to what we have seen, it could not be otherwise. He endeavoured to show that the nature-philosophers had no right to rear such comprehensive conclusions on the basis of the empirical knowledge then possessed, and that the unity of organization—or plan of structure of organisms—as maintained by them, did not exist. He represented the teleological (dualistic) conception of nature, and maintained that "the immutability of species was a necessary condition for the existence of a scientific history of nature," Cuvier had the great advantage over his opponent, that he was able to bring towards the proof of his assertions things obvious to the eye; these, however, were only individual facts taken out of their connection with others. Geoffroy was not able to prove the higher and general connection of individual phenomena which he maintained, by equally tangible details. Hence Cuvier, in the eyes of the majority, gained the victory, and decided the defeat of the nature-philosophy and the supremacy of the strictly empiric tendency for the next thirty years.

Goethe of course supported Geoffroy's views. How deeply interested he was, even in his 81st year, in this great contest is proved by the following anecdote related by Soret:—"Monday, Aug. 2nd, 1830.—The news of the outbreak of the revolution of July arrived in Weimar to-day, and has caused general excitement. In the course of the afternoon I went to Goethe. 'Well?' he exclaimed as I entered, 'what do you think of this great event? The volcano has burst forth, all is in flames, and there are no more negotiations behind closed doors.' 'A dreadful affair,' I answered; 'but what else could be expected under the circumstances, and with such a ministry, except that it would end in the expulsion of the present royal family?' 'We do not seem to understand each other, my dear friend,' replied Goethe. 'I am not speaking of those people at all; I am interested in something very different, I mean the dispute between Cuvier and Geoffroy de Saint Hilaire, which has broken out in the Academy, and which is of such great importance to science.' This remark of Goethe's came upon me so unexpectedly, that I did not know what to say, and my thoughts for some minutes seemed to have come to a complete standstill. 'The affair is of the utmost importance,' he continued, 'and you cannot form any idea of what I felt on receiving the news of the meeting on the 19th. In Geoffroy de Saint Hilaire we have now a mighty ally for a long time to come. But I see also how great the sympathy of the French scientific world must be in this affair, for, in spite of the terrible political excitement, the meeting on the 19th was attended by a full house. The best of it is, however, that the synthetic treatment of nature, introduced into France by Geoffroy, can now no longer be stopped. This matter has now become public through the discussions in the Academy, carried on in the presence of a large audience; it can no longer be referred to secret committees, or be settled or suppressed behind closed doors.'"

In my book on "The General Morphology of Organisms" I have placed as headings to the different books and chapters a selection of the numerous interesting and important sentences in which Goethe clearly expresses his view of organic nature and its constant development. I will here quote a passage from the poem entitled, "The Metamorphosis of Animals" (1819).

> "All members develop themselves according to eternal laws,
> And the rarest form mysteriously preserves the primitive type,
> Form therefore determines the animal's way of life,
> And in turn the way of life powerfully reacts upon all form.
> Thus the orderly growth of form is seen to hold
> Whilst yielding to change from externally acting causes."[*]

Here, clearly enough, the contrast between two different organic constructive forms is intimated, which are opposed to one another, and which by their interaction determine the form of the organism; on the one hand, a common inner original type, firmly maintaining itself, constitutes the foundation of the most different forms; on the other hand, the externally active influence of surroundings and mode of life, which influence the original type and transform it. This contrast is still more definitely pointed out in the following passage:—

"An inner original community forms the foundation of all organization; the variety of forms, on the other hand, arises from the necessary relations

---

[*] Alle Glieder bilden sich aus nach ew'gen Gesetzen,
Und die seltenste Form bewahrt im Geheimniss das Urbild.
Also bestimmt die Gestalt die Lebensweise des Thieres.
Und die Weise zu leben, sie wirkt auf alle Gestalten
Mächtig züruck. So zeiget sich fest die geordnete Bildung,
Welche zum Wechsel sich neight durch äusserlich wirkende Wesen.

to the outer world, and we may therefore justly assume an original difference of conditions, together with an uninterruptedly progressive transformation, in order to be able to comprehend the constancy as well as the variations of the phenomena of form."

The "original type" which constitutes the foundation of every organic
form "as the inner original community" is the *inner constructive force*, which
receives the original direction of form-production—that is, the tendency to
give rise to a particular form—and is propagated by *Inheritance*. The "uninterruptedly progressive transformation," on the other hand, which "springs
from the necessary relations to the outer world," acting as an *external formative force*, produces, by *Adaptation* to the surrounding conditions of life,
the "infinite variety of forms" (Gen. Morph. i. 154; ii. 224). The internal formative tendency of *Inheritance*, which retains the unity of the original type,
is called by Goethe in another passage the *centripetal force* of the organism,
or its tendency to specification; in contrast with this he calls the external
formative tendency of *Adaptation*, which produces the variety of organic
forms, the *centrifugal force* of organisms, or their tendency to variation.
The passage in which he clearly indicates the "equilibrium" of these two extremely important organic formative tendencies, runs as follows: "The idea
of *metamorphosis* resembles the vis centrifuga, and would lose itself in the
infinite, if a counterpoise were not added to it: I mean the tendency to *specification*, the strong power to preserve what once has come into being, a vis
centripeta, which in its deepest foundation cannot be affected by anything
external."

Metamorphosis, according to Goethe, consists not merely, as the word is
now generally understood, in the changes of form which the organic individual experiences during its individual development, but, in a wider sense, in
the transformation of organic forms in general. His idea of metamorphosis
is almost synonymous with the theory of development. This is clear, among
other things, from the following passage:—"The triumph of physiological
metamorphosis manifests itself where the whole separates and transforms
itself into families, the families into genera, the genera into species, and
then again into other varieties down to the individual. This operation of
nature goes on ad infinitum; she cannot rest inactive, but neither can she
keep and preserve all that she has produced. From seeds there are always
developed varying plants, exhibiting the relations of their parts to one another in an altered manner."

Goethe had, in truth, discovered two great mechanical forces of nature,
which are the active causes of organic formations, his two organic formative
tendencies—on the one hand the conservative, centripetal, and internal formative tendency of Inheritance or specification; and on the other hand the
progressive, centrifugal, and external formative tendency of Adaptation, or
metamorphosis. This profound biological intuition could not but lead him
naturally to the fundamental idea of the Doctrine of Filiation, that is, to the

conception that the organic species resembling one another in form are actually related by blood, and that they are descended from a common original type. In regard to the most important of all animal groups, namely that of Vertebrate animals, Goethe expresses this doctrine in the following passage (1796):—"Thus much then we have gained, that we may assert without hesitation that all the more perfect organic natures, such as fishes, amphibious animals, birds, mammals, and man at the head of the last, were all formed upon one original type, which only varies more or less in parts which are none the less permanent, and still daily changes and modifies its form by propagation."

This sentence is of interest in more than one way. The theory that all "the more perfect organic natures," that is all Vertebrate animals, are descended from one common prototype, that they have arisen from it by propagation (Inheritance) and transformation (Adaptation), may be distinctly inferred. But it is especially interesting to observe that Goethe admits no exceptional position for man, but rather expressly includes him in the tribe of the other Vertebrate animals. The most important special inference of the Doctrine of Filiation, that man is descended from other Vertebrate animals, may here be recognized in the germ.[3]

This exceedingly important fundamental idea is expressed by Goethe still more clearly in another passage (1807), in the following words:—"If we consider plants and animals in their most imperfect condition, they can scarcely be distinguished. But this much we can say, that the creatures which by degrees emerge as plants and animals out of a common phase, where they are barely distinguishable, arrive at perfection in two opposite directions; so that the plant in the end reaches its highest glory in the tree, which is immovable and stiff, the animal in man, who possesses the greatest elasticity and freedom." This remarkable passage not only indicates most explicitly the genealogical relationship between the vegetable and animal kingdoms, but contains the germ of the monophyletic hypothesis of descent, the importance of which I shall have to explain hereafter. (Compare Chapter XVI. and the Pedigree, p. 398.)

At the time when Goethe in this way sketched the fundamental features of the Theory of Descent, another German philosopher, Gottfried Reinhold Treviranus, of Bremen (born 1776, died 1837), was zealously engaged at the same work. As Wilhelm Focke has recently shown, Treviranus, even in the earliest of his greater works, "The Biology or Philosophy of Animate Nature," which appeared at the beginning of the present century, had already developed monistic views of the unity of nature, and of the genealogical connection of the species of organisms, which entirely correspond with our present view of the matter. In the first three volumes of the Biology, which appeared successively in 1802, 1803, and 1805 (therefore several years before Oken's and Lamarck's principal works), we find numerous passages which are of interest in this respect. I shall here quote only a few of the most important.

In speaking of the principal question of our theory, the question of the origin of organic species, Treviranus makes the following remarks:—"Every form of life can be produced by physical forces in one of two ways: either by coming into being out of formless matter, or by modification of an already existing form by a continued process of shaping. In the latter case the cause of this modification may lie either in the influence of a dissimilar male generative matter upon the female germ, or in the influence of other powers which operate only after procreation. In every living being there exists the capability of an endless variety of form-assumption; each possesses the power to adapt its organization to the changes of the outer world, and it is this power put into action by the change of the universe that has raised the simple zoophytes of the primitive world to continually higher stages of organization, and has introduced a countless variety of species into animate nature."

By *zoophytes*, Treviranus here means organisms of the lowest order and of the simplest character, namely, those neutral primitive beings which stand midway between animals and plants, and on the whole correspond with our *protista*. "These zoophytes," he remarks in another passage, "are the original forms out of which all the organisms of the higher classes have arisen by gradual development. We are further of opinion that every species, as well as every individual, has certain periods of growth, of bloom, and of decay, but that the decay of a species is *degeneration*, not dissolution, as in the case of the individual. From this it appears to us to follow that it was not the great catastrophes of the earth (as is generally supposed) which destroyed the animals of the primitive world, but that many survived them, and it is more probable that they have disappeared from existing nature, because the species to which they belonged have completed the circle of their existence, and have become changed into other kinds."

When Treviranus, in this and other passages, points to *degeneration* as the most important cause of the transformation of the animal and vegetable species, he does not understand by it what is now commonly called degeneration. With him "degeneration" is exactly what we now call *Adaptation* or *modification*, by the action of external formative forces. That Treviranus explained this trans-transformation of organic species by Adaptation, and its preservation by Inheritance, and thus the whole variety of organic forms by the interaction of Adaptation and Inheritance, is clear also from several other passages. How profoundly he grasped the mutual dependence of all living creatures on one another, and in general the *universal connection between cause and effect*—that is, the monistic causal connection between all members and parts of the universe—is further shown, among others, by the following remarks in his Biology:—"The living individual is dependent upon the species, the species upon the fauna, the fauna upon the whole of animate nature, and the latter upon the organism of the earth. The individual possesses indeed a peculiar life, and so far forms its own world. But just because its life is limited it constitutes at the same time an organ in the general organism. Every living body exists in consequence of the universe, but the universe, on the other hand, exists in consequence of it."

It is self-evident that so profound and clear a thinker as Treviranus, in accordance with this grand mechanical conception of the universe, could not admit for man a privileged and exceptional position in nature, but assumed his gradual development from lower animal forms. And it is equally self-evident, on the other hand, that he did not admit a chasm between organic and inorganic nature, but maintained the absolute unity of the organization of the whole universe. This is specially attested by the following sentence:— "Every inquiry into the influence of the whole of nature on the living world must start from the principle, that all living forms are products of physical influences, which are acting even now, and are changed only in degree, or in their direction." Hereby, as Treviranus himself says, "The fundamental problem of biology is solved," and we add, solved in a purely mechanical or monistic sense.

Neither Treviranus nor Goethe is commonly considered the most eminent of the German nature-philosophers, but Lorenz Oken, who, in establishing the vertebral theory of the skull, came forward as a rival to Goethe, and did not entertain a very kindly feeling towards him. Although they lived for some time in the same neighbourhood, yet the natures of these two men were so very different, that they could not well be drawn towards each other. Oken's "Manual of the Philosophy of Nature," which may be designated as the most important production of the nature-philosophy school then existing in Germany, appeared in 1809, the same year in which Lamarck's fundamental work, the "Philosophie Zoologique," was published. As early as 1802, Oken had published an "Outline of the Philosophy of Nature." As we have already intimated, in Oken's as in Goethe's works, a number of valuable and profound thoughts are hidden among a mass of erroneous, very eccentric, and fantastic conceptions. Some of these ideas have only quite recently and gradually become recognized in science, many years after they were first expressed. I shall here quote only two thoughts, which are almost prophetic, and which at the same time stand in the closest relation to the theory of development.

One of the most important of Oken's theories, which was formerly very much decried, and was most strongly combatted, especially by the so-called "exact experimentalists," is the idea that the phenomena of life in all organisms proceed from a common chemical substance, so to say, from a general simple *vital-substance*, which he designated by the name *Urschleim*, or *original slime*. By it he meant, as the name indicates, a mucilaginous substance, an albuminous combination, which exists in a semi-fluid condition of aggregation, and possesses the power, by adaptation to different conditions of existence in the outer world and by interaction with its material, of producing the most various forms. Now, we need only change the expression "original slime" (Urschleim) into *Protoplasm*, or *cell-substance*, in order to

arrive at one of the grandest results which we owe to microscopic investigations during the last ten years, more especially to those of Max Schultze. By
these investigations it has been shown that in all living bodies, without exception, there exists a certain quantity of mucilaginous albuminous matter,
in a semi-fluid condition; and that this nitrogen-holding carbon-compound
is exclusively the original seat and agent of all the phenomena of life, and
of all production of organic forms. All other substances which appear in the
organism, besides these, are either formed by this active matter of life, or
have been introduced from without. The organic egg, the original cell out of
which every animal and plant is first developed, consists essentially only of
one round little lump of such albuminous matter. Even the yolk of an egg is
nothing but albumen, mixed with granules of fat. Oken was therefore right
when, more divining than knowing, he made the assertion—"Every organic
thing has arisen out of slime, and is nothing but slime in different forms.
This primitive slime originated in the sea, from inorganic matter in the
course of planetary-evolution."

Another equally grand idea of the same philosopher is closely connected
with his theory of primitive slime, which coincides with the extremely important *Protoplasm theory*. For Oken, as early as 1809, asserted that the
primitive slime produced in the sea by spontaneous generation, at once assumed the form of microscopically small bladders, which he called "*Mile*,"
or "*Infusoria*." "Organic nature has for its basis an infinity of such vesicles."
These little bladders arise from original semi-fluid globules of the primitive
slime, by the fact of their periphery becoming condensed. The simplest organism, as well as every animal and every plant of higher kind, is nothing
else than "an accumulation (synthesis) of such infusorial bladders, which by
various combinations assume various forms, and thus develop into higher
organisms." Here again we need only translate the expression *little bladder*,
or *infusorium*, by the word *cell*, and we arrive at the Cell theory, one of the
grandest biological theories of our century. Schleiden and Schwann, about
thirty years ago, were the first to furnish experiential proof that all organisms are either simple cells, or accumulations (syntheses) of such cells, and
the more recent protoplasm theory has shown that protoplasm (the original
slime) is the most essential (and sometimes the only) constituent part of the
genuine cell. The properties which Oken ascribes to his Infusoria are exactly the properties of cells, the properties of elementary beings, by whose accumulation, combination, and varying development, the higher organisms
are formed.

These two extremely fruitful thoughts of Oken, on account of the absurd
form in which he expressed them, were at first little heeded, or entirely
misunderstood, and it was reserved for a much later era to establish them
by actual observation. The supposition that the individual species of plants
and animals originated from common prototypes by a slow and gradual development of the higher organisms out of lower ones, was of course most

closely connected with these ideas. Man's descent from lower organisms was likewise asserted by Oken—"Man has been developed, not created." Although many arbitrary perversities and extravagant fancies may be found in Oken's philosophy of nature, they must not prevent us paying our just admiration to these grand ideas, which were so far in advance of their age. This much is clearly evident from the statements of Goethe and Oken which we have quoted, and from the views of Lamarck and Geoffroy which have to be discussed next, that during the first decade of our century no doctrine approached so nearly to the natural Theory of Descent, newly established by Darwin, as the much decried "Natur-philosophie."

CHAPTER V.

THEORY OF DEVELOPMENT ACCORDING TO KANT AND LAMARCK.

Kant's Dualistic Biology.—His Conception of the Origin of Inorganic Nature by Mechanical Causes, of Organic Nature by Causes acting for a Definite Purpose.—Contradiction of this Conception with his leaning towards the Theory of Descent.—Kant's Genealogical Theory of Development.—Its Limitation by his Teleology.—Comparison of Genealogical Biology with Comparative Philology.—Views in favour of the Theory of Descent entertained by Leopold Buch, Bär, Schleiden, Unger, Schaafhausen, Victor Carus, Büchner.—French Nature-philosophy.—Lamarck's Philosophie Zoologique.—Lamarck's Monistic (mechanical) System of Nature.—His Views of the Interaction of the Two Organic Formative Tendencies of Inheritance and Adaptation.—Lamarck's Conception of Man's Development from Ape-like Mammals.—Geoffroy St. Hilaire's, Naudin's, and Lecoq's Defence of the Theory of Descent.—English Nature-philosophy.—Views in favour of the Theory of Descent, entertained by Erasmus Darwin, W. Herbert, Grant, Freke, Herbert Spencer, Hooker, Huxley.—The Double Merit of Charles Darwin.

THE teleological view of nature, which explains the phenomena of the organic world by the action of a personal Creator acting for a definite purpose, necessarily leads, when carried to its extreme consequences, either to utterly untenable contradictions, or to a twofold (dualistic) conception of nature, which most directly contradicts the unity and simplicity of the supreme laws which are everywhere perceptible. The philosophers who embrace teleology must necessarily assume two fundamentally different natures: an *inorganic* nature, which must be explained by causes acting *mechanically* (causae efficientes), and an *organic* nature, which must be explained by *causes acting for a definite purpose* (causae finales). (Compare p. 34. [19])

This dualism meets us in a striking manner when considering the conceptions of nature formed by Kant, one of the greatest German philosophers, and his ideas of the coming into being of organisms. A closer examination of these ideas is forced upon us here, because in Kant we honour one of the few philosophers who combine a solid scientific culture with an extraordinary clearness and profundity of speculation. The Königsberg philosopher gained the highest celebrity, not only among speculative philosophers as the founder of critical philosophy, but acquired a brilliant name also among naturalists by his mechanical cosmogeny. Even in the year 1755, in his "General History of Nature, and Theory of the Heavens,"[22] he made the bold attempt "to discuss the constitution and the mechanical origin of the whole universe, according to Newton's principles," and to explain them mechanically by the natural course of development, to the exclusion of all miracles. This cosmogeny of Kant, or "cosmological gas theory," which we shall briefly discuss in a future chapter, was at a later day fully established by the French mathematician Laplace and the English astronomer Herschel, and enjoys at the present day almost universal recognition. On account of this important work alone, in which exact knowledge is coupled with most profound speculation, Kant deserves the honourable name of a natural philosopher in the best and purest sense of the word.

If we read Kant's Criticism of the Teleological Faculty of Judgment, his most important biological work, we perceive that in contemplating organic nature he always maintains what is essentially the teleological or dualistic point of view; whilst for inorganic nature he, unconditionally and without reserve, assumes the mechanical or monistic method of explanation. He affirms that in the domain of inorganic nature all the phenomena can be explained by mechanical causes, by the moving forces of matter itself, but not so in the domain of organic nature. In the whole of Anorganology (in Geology and Mineralogy, in Meteorology and Astronomy, in the physics and chemistry of inorganic natural bodies), all phenomena are said to be explicable merely by *mechanism* (causa efficiens), without the intervention of a final purpose. In the whole domain of Biology, on the other hand—in Botany, Zoology, and Anthropology—mechanism is not considered sufficient to explain to us all their phenomena; but we are supposed to be able to comprehend them only by an assumption of a *final cause* acting for a definite purpose (causa finalis). In several passages Kant emphatically remarks that, from a strictly scientific point of view, *all* phenomena, without exception, require a mechanical interpretation, and that *mechanism alone can offer a true explanation*. But at the same time he thinks, that in regard to living natural bodies, animals and plants, our human power of comprehension is limited, and not sufficient for arriving at the real cause of organic processes, especially at the origin of organic forms. The *right* of human reason to explain all phenomena mechanically is unlimited, he says, but its *power* is limited by the fact that organic nature can be conceived only from a teleological point of view.

Some passages are, however, very remarkable, in which Kant in a surprising manner deviates from this mode of viewing things, and expresses, more or less distinctly, the fundamental idea of the Theory of Descent. He even asserts the necessity of a genealogical conception of the series of organisms, if we at all wish to understand it scientifically. The most important and remarkable of these passages occurs in his "Methodical System of the Teleological Faculty of Judgment" (Sec. 79), which appeared in 1790 in the

"Criticism of the Faculty of Judgment." Considering the extraordinary interest which this passage possesses, both for forming a correct estimate of Kant's philosophy, as well as for the Theory of Descent, I shall here insert it *verbatim.*

"It is desirable to examine the great domain of organized nature by means of a methodical comparative anatomy, in order to discover whether we may not find in it something resembling a system, and that too in connection with the mode of generation, so that we may no longer be compelled to stop short with a mere consideration of forms as they are—which gives us no insight into their generation—and need no longer give up in despair all hope of gaining a full insight into this department of nature. The agreement of so many kinds of animals in a certain common plan of structure, which seems to be visible not only in their skeletons, but also in the arrangement of the remaining parts—so that a wonderfully simple typical form, by the shortening and lengthening of some parts, and by the suppression and development of others, might be able to produce an immense variety of species—gives us a ray of hope, though feeble, that here perhaps some result may be obtained, by the application of the principle of the *mechanism of nature*, without which, in fact, no science can exist. This analogy of forms (in so far as they seem to have been produced in accordance with a common prototype, notwithstanding their great variety) strengthens the supposition that they have an actual blood-relationship, due to origination from a common parent; a supposition which is arrived at by observation of the graduated approximation of one class of animals to another, beginning with the one in which the principle of purposiveness seems to be most conspicuous, that is man, and extending down to the polyps, and from these even down to mosses and lichens, and arriving finally at raw matter, the lowest stage of nature observable by us. From this matter and its forces the whole apparatus of Nature seems to have descended according to mechanical laws (such as those which she follows in the production of crystals); yet this apparatus, as seen in organic beings, is so incomprehensible to us, that we feel ourselves compelled to conceive for it a different principle. But it would seem that the archaeologist of Nature is at liberty to regard the great *Family* of creatures (for as a Family we must conceive it, if the above-mentioned continuous and connected relationship has a real foundation) as having sprung from the immediate results of her earliest revolutions, judging from all the laws of their mechanism known to or conjectured by him."

If we take this remarkable passage out of Kant's "Criticism of the Teleological Faculty of Judgment," and consider it by itself, we cannot but be astonished to find how profoundly and clearly the great thinker, even in 1790, had recognized the inevitable necessity of the Doctrine of Descent, and designated it as the only possible way of explaining organic nature by mechanical laws—that is, by true scientific reasoning. On account of this one passage taken by itself, we might place Kant beside Goethe and Lamarck, as one of the first founders of the Doctrine of Descent; and considering the high authority which Kant's Critical Philosophy most justly enjoys, this circumstance might perhaps induce many a philosopher to decide in favour of the theory. But as soon as we consider this passage in connection with the other train of thoughts in the "Criticism of the Faculty of Judgment," and balance it against other directly contradictory passages, we see clearly that Kant, in these and some similar (but weaker) sentences, went beyond himself, and abandoned the teleological point of view which he usually adopts in Biology.

Directly after the admirable passage which I have just quoted, there follows a remark which completely takes off its edge. After having quite correctly maintained the origin of organic forms out of raw matter by mechanical laws (in the manner of crystallization), as well as a gradual development of the different species by descent from one common original parent, Kant adds, "But he (the archaeologist of nature, that is the palaeontologist) must for this end ascribe to the common mother an organization ordained purposely with a view to the needs of all her offspring, otherwise the possibility of suitability of form in the products of the animal and vegetable kingdoms (*i.e.* teleological adaptation) cannot be conceived at all." This addition clearly contradicts the most important fundamental thought of the preceding passage, viz., that a purely mechanical explanation of organic nature becomes possible through the Theory of Descent. And that the teleological conception of organic nature predominated with Kant, is shown by the heading of the remarkable Sec. 79, which contains the two contradictory passages cited: *"Of the Necessary Subordination of the Mechanical to the Teleological Principle, in the explanation of a thing as a purpose or object of Nature."*

He expresses himself most decidedly against the mechanical explanation of organic nature in the following passage (Sec. 74): "It is quite certain that we cannot become sufficiently acquainted with organized creatures and their hidden potentialities by aid of purely mechanical natural principles, much less can we explain them; and this is so certain, that we may boldly assert that it is absurd for man even to conceive such an idea, or to hope that a Newton may one day arise able to make the production of a blade of grass comprehensible, according to natural laws ordained by no intention; such an insight we must absolutely deny to man." Now, however, this impossible Newton has really appeared seventy years later in Darwin, whose Theory of Selection has actually solved the problem, the solution of which Kant had considered absolutely inconceivable!

In connection with Kant and the German philosophers whose theories of development have already occupied us in the preceding chapter, it seems justifiable to consider briefly some other German naturalists and philosophers, who, in the course of our century, have more or less distinctly resisted the prevailing teleological views of creation, and vindicated the mechanical conception of things which is the basis of the Doctrine of

Filiation. Sometimes general philosophical considerations, sometimes special empirical observations, were the motives which led these thinking men to form the idea that the various individual species of organisms must have originated from common primary forms. Among them I must first mention the great German geologist, Leopold Buch. Important observations as to the geographical distribution of plants led him to the following remarkable assertion in his excellent "Physical Description of the Canary Islands":—

"The individuals of genera, on continents, spread and widely diffuse themselves, and by the difference of localities, nourishment, and soil, form varieties; and being in consequence of their isolation never crossed by other varieties, and so brought back to the main type, they in the end become a permanent and a distinct species. Then, perhaps, in other ways, they once more become associated with other descendants of the original form—which have likewise become new varieties—and both now appear as very distinct species, no longer mingling with one another. Not so on islands. Being commonly confined in narrow valleys or within the limit of small zones, individuals can reach one another and destroy every commencing production of a permanent variety. Much in the same way the peculiarities or faults in language, originating with the head of some family, become, through the extension of the family, indigenous throughout a whole district. If the district is separated and isolated, and if the language is not brought back to its former purity by constant connection with that spoken in neighbouring districts, a dialect will be the result. If natural obstacles, forests, constitution, form of government, unite the inhabitants of the separate district still more closely, and separate them still more completely from their neighbours, the dialect is fixed, and becomes a completely distinct language." (Übersicht der Flora auf den Canarien, S. 133.)

We perceive that Buch is here led to the fundamental idea of the Theory of Descent by the phenomena of the geography of plants, a department of biological knowledge which in fact furnishes a mass of proofs in favour of it. Darwin has minutely discussed these proofs in two separate chapters of his book (the 11th and 12th). Buch's remark is further of interest, because it leads us to the exceedingly instructive comparison of the different branches of language with the species of organisms, a comparison which is of the greatest use to Comparative Philology, as well as to Comparative Botany and Zoology. Just as, for example, the different dialects, provincialisms, branches, and off-shoots of the German, Slavonic, Greco-Latin, and Irano-Indian parent language, are derived from a single common Indo-Germanic parent tongue, and just as their *differences* are explained by *Adaptation*, and their common *fundamental characters* explained by *Inheritance*, so in like manner the different species, genera, families, orders, and classes of Vertebrate animals are derived from a single common vertebrate form of animal. Here also Adaptation is the cause of differences, Inheritance the cause of community of character. This interesting parallelism in the divergent development of the forms of speech and the forms of organisms has been discussed in the clearest manner by one of our first comparative philologists, the talented Augustus Schleicher, whose premature death, four years ago, remains an irreparable loss, not only to our University of Jena, but to the whole of monistic science.[6]

Among other eminent German naturalists who have expressed their belief in the Theory of Descent more or less distinctly, arriving at their conclusion in very various ways, I must next mention Carl Ernst Bär, the great reformer of animal embryology. In a lecture delivered in 1834, entitled "The Most General Laws of Nature in All Development," he shows, in the clearest way, that only in a very childish view of nature could organic species be regarded as permanent and unchangeable types, and that really they can be only passing series of generations, which have developed by transformation from a common original form. The same conception again received firm support from Bär, in 1859, through a consideration of the laws of the geographical distribution of organisms.

J. M. Schleiden, who founded, thirty years ago, in Jena, a new epoch in Botany by his strictly empirico-philosophical and truly scientific method, illustrated the philosophical significance of the conception of organic species in his incisive "Outlines of Scientific Botany,"[7] and showed that it had only a subjective origin in the general *law of specification.* The different species of plants are only the specified productions of the formative tendencies of plants, which arise from the various combinations of the fundamental forces of organic matter.

The eminent botanist, F. Unger, of Vienna, was led by his profound and comprehensive investigations on extinct vegetable species, to a palaeontological history of the development of the vegetable kingdom, which distinctly asserts the principle of the Theory of Descent. In his "Attempt at a History of the World of Plants" (1852), he maintains the derivation of all different species of plants from a few primary forms, and perhaps from a single original plant, a simple vegetable cell. He shows that this view is founded on the genetic connection of all vegetable forms, and is necessary, not merely upon philosophical grounds, but upon those of experience and observation.[8]

Victor Carus, of Leipzig, in the Introduction to his excellent "System of Animal Morphology,"[9] published in 1853, in which he endeavours to establish in a philosophical manner the universal constructive laws of the animal body through comparative anatomy and the history of development, makes the following remark:—"The organisms buried in the most ancient geological strata must be looked upon as the ancestors from whom the rich diversity of forms of the present creation have originated by continued generation, and by accommodation to progressive and very different conditions of life."

In the same year (1853) Schaaffhausen, the anthropologist of Bonn, in an Essay "On the Permanence and Transformation of Species," declared himself decidedly in favour of the Theory of Descent. According to him, the living species of animals and plants are the transformed descendants of extinct species, from which they have arisen by gradual modification. The divergence or separation of the most nearly allied species takes place by the destruction of the connecting intermediate stages. Schaaffhausen also maintained, with distinctness, the origin of the human race from animals, and its gradual development from ape-like animals, the most important deduction from the Doctrine of Filiation.

Lastly, we have still to mention among the German Nature-philosophers the name of Louis Büchner, who, in his celebrated work, "Force and Matter" (1855), also independently developed the principles of the Theory of Descent, taking his stand mainly on the ground of the undeniable evidences of fact which are furnished by the palaeontological and individual development of organisms, as well as by their comparative anatomy and by the parallelism of these series of development. Büchner showed very clearly that, even from such data alone, the derivation of the different organic species from common primary forms followed as a necessary conclusion, and that the origin of these original primary forms could only be conceived of as the result of a spontaneous generation.

We now turn from the German to the French Nature-philosophers, who have likewise held the Theory of Descent, since the beginning of the present century. At their head stands Jean Lamarck, who occupies the first place next to Darwin and Goethe in the history of the Doctrine of Filiation. To him will always belong the immortal glory of having for the first time worked out the Theory of Descent, as an independent scientific theory of the first order, and as the philosophical foundation of the whole science of Biology. Although Lamarck was born as early as 1744, he did not begin the publication of his theory until the commencement of the present century, in 1801, and established it more fully only in 1809, in his classic "Philosophie Zoologique."[2] This admirable work is the first connected exposition of the Theory of Descent carried out strictly into all its consequences. By its purely mechanical method of viewing organic nature, and the strictly philosophical proofs brought forward in it, Lamarck's work is raised far above the prevailing dualistic views of his time; and with the exception of Darwin's work, which appeared just half a century later, we know of none which we could in this respect place by the side of the "Philosophie Zoologique." How far it was in advance of its time is perhaps best seen from the circumstance that it was not understood by most men, and for fifty years was not spoken of at all. Cuvier, Lamarck's greatest opponent, in his "Report on the Progress of Natural Sciences," in which the most unimportant anatomical investigations are enumerated, does not devote a single word to this work, which forms an epoch in science. Goethe, also, who took such a lively interest in the French nature-philosophy and in "the thoughts of kindred minds beyond the Rhine," nowhere mentions Lamarck, and does not seem to have known the "Philosophie Zoologique" at all. The great reputation which Lamarck gained as a naturalist he does not owe to his highly important general work, but to numerous special treatises on the lower animals, particularly on Molluscs, as well as to an excellent "Natural History of Invertebrate Animals," which appeared, in seven volumes, between the years 1815–1822. The first volume of this celebrated work contains in the general introduction a detailed exposition of his theory of filiation. I can, perhaps, give no better idea of the extraordinary importance of the "Philosophie Zoologique" than by quoting *verbatim* some of the most important passages therefrom:—

"The systematic divisions of classes, orders, families, genera, and species, as well as their designations, are the arbitrary and artificial productions of man. The kinds or species of organisms are of unequal age, developed one after the other, and show only a relative and temporary persistence; species arise out of varieties. The differences in the conditions of life have a modifying influence on the organization, the general form, and the parts of animals, and so has the use or disuse of organs. In the first beginning only the very simplest and lowest animals and plants came into existence; those of a more complex organization only at a later period. The course of the earth's development, and that of its organic inhabitants, was continuous, not interrupted by violent revolutions. Life is purely a physical phenomenon. All the phenomena of life depend on mechanical, physical, and chemical causes, which are inherent in the nature of matter itself. The simplest animals and the simplest plants, which stand at the lowest point in the scale of organization, have originated and still originate by spontaneous generation. All animate natural bodies or organisms are subject to the same laws as inanimate natural bodies or anorgana. The ideas and actions of the understanding are the motional phenomena of the central nervous system. The will is in truth never free. Reason is only a higher degree of development and combination of judgments."

These are indeed astonishingly bold, grand, and far-reaching views, and were expressed by Lamarck sixty years ago; in fact, at a time when their establishment, by a mass of facts, was not nearly as possible as it is in our day. Indeed Lamarck's work is really a complete and strictly monistic (mechanical) system of nature, and all the important general principles of monistic Biology are already enunciated by him: the unity of the active causes in organic and inorganic nature; the ultimate explanation of these causes in the chemical and physical properties of matter itself; the absence of a special vital power, or of an organic final cause; the derivation of all organisms from some few, most simple original forms, which have come into existence by spontaneous generation out of inorganic matter; the coherent course of the whole earth's history; the absence of violent cataclysmic revolutions; and in general the inconceivableness of any miracle, of any supernatural interference, in the natural course of the development of matter.

The fact that Lamarck's wonderful intellectual feat met with scarcely any recognition, arises partly from the immense length of the gigantic stride with which he had advanced beyond the next fifty years, partly from its defective empirical foundation, and from the somewhat one-sided character of some of his arguments. Lamarck quite correctly recognizes *Adaptation* as the first mechanical cause which effects the continual transformation of organic forms, while he traces with equal justice the similarity in form of different species, genera, families, etc., to their blood-relationship, and thus explains it by *Inheritance.* Adaptation, according to him, consists in this, that the perpetual, slow change of the outer world causes a corresponding change in the actions of organisms, and thereby also causes a further change in their forms. He lays the greatest stress upon the effect of *habit* upon the use and disuse of organs. This is certainly of great importance in the transformation of organic forms, as we shall see later. However, the way in which Lamarck wished to explain exclusively, or at any rate mainly, the change of forms, is after all in most cases not possible. He says, for example, that the long neck of the giraffe has arisen from its constantly stretching out its neck at high trees, and from the endeavour to pick the leaves off their branches; as giraffes generally inhabit dry districts, where only the foliage of trees afford them nourishment, they were forced to this action. In like manner the long tongues of wood-peckers, humming-birds, and ant-eaters, are said by him to have arisen from the habit of fetching their food out of narrow, small, and deep crevices or channels. The webs between the toes of the webbed feet in frogs and other aquatic animals have arisen solely from the constant endeavour to swim, from striking their feet against the water, and from the very movements of swimming. Inheritance fixed these habits on the descendants, and finally, by further elaboration, the organs were entirely transformed. However correct, as a whole, this fundamental thought may be, yet Lamarck lays the stress too exclusively on *habit* (use and non-use of organs), certainly one of the most important, but not the only cause of the change of forms. Still this cannot prevent our acknowledging that Lamarck quite correctly appreciated the mutual co-operation of the two organic formative tendencies of Adaptation and Inheritance. What he failed to grasp is the exceedingly important principle of "Natural Selection in the Struggle for Existence," with which Darwin, fifty years later, made us acquainted.

It still remains to be mentioned as a special merit of Lamarck, that he endeavoured to prove the *development of the human race* from other primitive, ape-like mammals. Here again it was, above all, to habit that he ascribed the transforming, the ennobling influence. He assumed that the lowest, original men had originated out of men-like apes, by the latter accustoming themselves to walk upright. The raising of the body, the constant effort to keep upright, in the first place led to a transformation of the limbs, to a stronger differentiation or separation of the fore and hinder extremities, which is justly considered one of the most essential distinctions between man and the ape. Behind, the calf of the leg and the flat soles of the feet were developed; in front, the arms and hands, for the purpose of seizing objects. The upright walk was then followed by a freer view over the surrounding objects, and led consequently to an important progress in mental development. Human apes thereby soon gained a great advantage over the other apes, and further, over surrounding organisms in general. In order to maintain the supremacy over them, they formed themselves into companies, and there arose, as in the case of all animals living in company, the desire of communicating to one another their desires and thoughts. Thus arose the necessity of language, which, consisting at first of rough and disjointed sounds, soon became more connected, developed, and articulate. The development of articulate speech now in turn became the strongest lever for a further progressive development of the organism, and above all, of the brain, and so ape-like men became gradually and slowly transformed into real men. In this way the actual descent of the lowest and rudest primitive men from the most highly developed apes was distinctly maintained by Lamarck, and supported by a series of the most important proofs.

The honour of being the chief French nature-philosopher is not usually assigned to Lamarck, but to Etienne Geoffroy St. Hilaire (the elder), born in 1771, the same in whom Goethe was especially interested, and with whom we have already become acquainted as Cuvier's most prominent opponent. He developed his ideas about the transformation of organic species as far back as the end of the last century, but published them only in the year 1828, and then in the following years, especially in 1830, defended them bravely against Cuvier. Geoffroy St. Hilaire in all essentials adopted Lamarck's Theory of Descent, yet he believed that the transformation of animal and vegetable species was less effected by the action of the organism itself (by habit, practice, use, or disuse of organs) than by the "monde ambiant," that is, by the continual change of the outer world, especially of the atmosphere. He conceives the organism as passive, in regard to the vital conditions of the outer world, while Lamarck, on the contrary, regards it as active. Geoffroy thinks, for example, that birds originated from lizard-like reptiles, simply by a diminution of the carbonic acid in the atmosphere, in consequence of which the breathing process became more animated and energetic through the increased proportion of oxygen in the atmosphere. Thus there arose a higher temperature of the blood, an increased activity of the nerves and muscles, and the scales of the reptiles became the feathers of the birds, etc. This conception is based upon a correct thought, but although the change of the atmosphere, as well as the change of every other external condition of existence, certainly effects directly or indirectly the transformation of the organism, yet this single cause is by itself too unimportant for such effects to be ascribed to it. It is even less important than practice and habit, upon which Lamarck lays too much stress. Geoffroy's chief merit consists in his having vindicated the monistic conception of nature, the unity of organic forms, and the deep genealogical connection of the different organic types in the face of Cuvier's powerful influence. I have already mentioned in the preceding chapter (pp. 87, 88 [49, 50]) the celebrated disputes between the two great opponents in the Academy of Paris, especially the fierce conflicts on the 22nd of February, and on the 19th of July, in which Goethe took so lively an interest. On that occasion Cuvier remained the acknowledged victor, and since that time very little, or rather nothing, more has been done in France to further the development of the Doctrine of Filiation, and complete the monistic theory of development. This is evidently to be ascribed principally to the repressive influence exercised by Cuvier's great authority. Even at the present day the majority of the French naturalists are the disciples and blind followers of Cuvier. In no civilized country of Europe has Darwin's doctrine had so little effect and been so little understood as in France, so that in the further course of our examination we need not take the French naturalists into consideration. At most, there are two distinguished botanists, among the recent French naturalists, whom we may mention as having ventured to express themselves in favour of the mutability and transformation of species. These two men are Naudin (1852) and Lecoq (1854).

Having discussed the early services of German and French nature-philosophy in establishing the doctrine of descent, we turn to the third great country of Europe, to free England, which during the last ten years has become the chief seat and starting-point for the further working out and definite establishment of the theory of development. Englishmen, who now take such an active part in every great scientific progress of humanity, and are the first to promote the eternal truths of natural science, at the beginning of the century took but little part in the continental nature-philosophy and its most important progress, the Theory of Descent. Almost the only earlier English naturalist whom we have here to mention is Erasmus Darwin, the grandfather of the reformer of the Theory of Descent. In 1795 he published, under the title of "Zoonomia," a scientific work in which he expresses views very similar to those of Goethe and Lamarck, without, however, then knowing anything about these two men. It is evident that the Theory of Descent at that time pervaded the intellectual atmosphere. Erasmus Darwin lays great stress upon the transformation of animal and vegetable species by their own vital action and by their becoming accustomed to changed conditions of existence, etc. Next, W. Herbert, in 1822, expressed the opinion that species of animals and plants are nothing but varieties which have become permanent. In like manner Grant, in Edinburgh, in 1826, declared that new species proceed from existing species by continued transformation. In 1841 Freke maintained that all organic beings must be descended from a single primitive type. In 1852 Herbert Spencer demonstrated minutely, and in a very clear and philosophic manner, the necessity of the Doctrine of Filiation, and established it more firmly in his excellent "Essays," which appeared in 1858, and in his "Principles of Biology," which was published at a later date. He has, at the same time, the great merit of having applied the theory of development to psychology, and of having shown that the emotional and intellectual faculties could only have been acquired by degrees and developed gradually. Lastly, we have to mention that in 1859 Huxley, the first of English zoologists, spoke of the Theory of Descent as the only hypothesis of creation reconcilable with scientific physiology. The same year produced the "Introduction to the Flora of Tasmania," in which Hooker, the celebrated English botanist, adopts the Theory of Descent, supporting it with important observations of his own.

All the naturalists and philosophers with whom we have become acquainted in this brief historical survey, as men adopting the Theory of Development, merely arrived at the conception that all the different species of animals and plants which at any time have lived, and still live, upon the earth, are the gradually changed and transformed descendants of one or some few original and very simple prototypes, which latter arose out of inorganic matter by spontaneous generation. But none of them succeeded in placing this fundamental element of the doctrine of descent in relation with some cause, nor in satisfactorily explaining the transformation of organic species by the true demonstration of its mechanical antecedents. Charles Darwin was the first who solved this most difficult problem, and this forms the broad gulf which separates him from his predecessors.

The special merit of Charles Darwin is, in my opinion, twofold: in the first place, the doctrine of descent, the fundamental idea of which was already clearly expressed by Goethe and Lamarck, has been developed by him much more comprehensively, has been traced much more minutely in all directions, and carried out much more strictly and connectedly than by any of his predecessors; and secondly, he has established a new theory, which reveals to us the natural causes of organic development, the acting causes (causae efficientes) of organic form-production, and of the changes and transformations of animal and vegetable species. This is the theory which we call the Theory of Selection, or more accurately, the Theory of Natural Selection (selectio naturalis).

When we reflect that (with the few exceptions above mentioned) the whole science of Biology, before Darwin's time, was elaborated in accordance with the opposite views, and that almost all zoologists and botanists regarded the absolute independence of organic species as a self-evident inference from the results of all study of forms, we shall certainly not lightly value the twofold merit of Darwin. The false doctrine of the constancy and independent creation of individual species had gained such high authority, was so generally recognized, and was, moreover, so much favoured by delusive appearances, accepted by superficial observation, that, indeed, no small degree of courage, strength, and intelligence was required to rise as a reformer against its omnipotence, and to dash to pieces the structure artificially erected upon it. But, in addition to this, Darwin added to Lamarck's and Goethe's doctrine of descent the new and highly important principle of "natural selection."

We must sharply distinguish the two points—though this is usually not done—first, Lamarck's Theory of Descent, which only asserts *that* all animal and vegetable species are descended from common, most simple, and spontaneously generated prototypes; and secondly, Darwin's Theory of Selection, which shows us *why* this progressive transformation of organic forms took place, and what causes, acting mechanically, effected the uninterrupted production of new forms, and the ever increasing variety of animals and plants.

Darwin's immortal merit cannot be justly estimated until a later period, when the Theory of Development, after overthrowing all other theories of creation, will be recognized as the supreme principle of explanation in Anthropology, and, consequently, in all other sciences. At present, while in the hot contest for truth the name of Darwin is the watchword to the advocates of the natural theory of development, his merits are inaccurately appreciated on both sides, for some persons overestimate them as much as others underestimate them.

His merit is overestimated when he is regarded as the founder of the Theory of Descent, or of the whole of the Theory of Development. We have seen from the historical sketch in this and the preceding chapters, that the Theory of Development, as such, is not new; all philosophers who have refused to be led captive by the blind dogma of a supernatural creation, have been compelled to assume a natural development. But the Theory of Descent constituting the specially biological part of the universal Theory of Development, had already been so clearly expressed by Lamarck, and carried out so fully by him to its most important consequences, that we must honour him as the real founder of it. Hence it is only the Theory of Selection, and not that of Descent, which may be called *Darwinism*; but this is in itself of so much importance, that its value can scarcely be overestimated.

Darwin's merit is naturally underestimated by all his opponents. But it is scarcely possible in this matter to point to scientific opponents, who are entitled by profound biological culture to pronounce an opinion. For among all the works opposed to Darwin and the Theory of Descent yet published, with the exception of that of Agassiz, not one deserves consideration, much less refutation; all have so evidently been written either without thorough knowledge of biological facts, or without a clear philosophical understanding of the question in hand. We need not trouble ourselves at all about the attacks of theologians and other unscientific men, who really know nothing whatever of nature.

The only eminent scientific adversary who still remains opposed to Darwin and the whole theory of development is Louis Agassiz; but the principle of his opposition in reality deserves notice only as a philosophical curiosity. In a French translation of his "Essay on Classification,"[5] which we have spoken of before, published in Paris in 1869, Agassiz has most formally announced his opposition to Darwinism, which he had previously expressed in many ways. To this translation he has appended a treatise of sixteen pages, bearing the title, "Le Darwinisme. Classification de Haeckel." This curious chapter contains the most wonderful things; as, for example, "Darwin's idea is a conception *à priori*. Darwinism is a burlesque of facts. Science would renounce the claim which it has hitherto possessed to the confidence of earnest minds if such sketches were to be accepted as indications of a true progress." The following passage, however, is the climax of this strange polemic: "Darwinism shuts out almost the whole mass of acquired knowledge in order to retain and assimilate to itself that only which may serve its doctrine."

Surely this is what we may call turning the whole affair topsy-turvy! The biologist who knows the facts must be astounded at Agassiz's courage in uttering such sentences—sentences without a word of truth in them, and which he cannot himself believe! The impregnable strength of the Theory of Descent lies just in the fact that all biological facts are explicable only through it, and that without it they remain unintelligible miracles. All our "laborious knowledge" in comparative anatomy and physiology—in embryology and palaeontology—in the doctrine of the geographical and topographical distribution of organisms, etc., constitutes an irrefutable testimony to the truth of the Theory of Descent.

In my General Morphology, especially in the sixth book (in the General Phylogeny), I have minutely refuted Agassiz's "Essay on Classification" in all essential points. The twenty-fourth chapter I have devoted to a very detailed and strictly scientific discussion of that section which Agassiz himself considers the most important (the groups or categories of systematic zoology and botany), and have shown that this part of his work is purely chimerical, without any trace of real foundation. Agassiz takes good care not to venture anywhere to touch upon my refutation, because, forsooth, he is not in a position to produce anything substantial against it. He fights not with arguments, but with phrases. However, such opposition will not delay the complete victory of the Theory of Development, but only accelerate it.

CHAPTER VI.

THEORY OF DEVELOPMENT ACCORDING TO
LYELL AND DARWIN.

Charles Lyell's Principles of Geology.—His Natural History of the Earth's Development.— Origin of the Greatest Effects through the Multiplication of the Smallest Causes.— Unlimited Extent of Geological Periods.—Lyell's Refutation of Cuvier's History of Creation.—The Establishment of the Uninterrupted Connection of Historical Development by Lyell and Darwin.—Biographical Notice of Charles Darwin.—His Scientific Works.—His Theory of Coral Reefs.—Development of the Theory of Selection.—A Letter of Darwin's.—The Contemporaneous Appearance of Darwin's and Alfred Wallace's Theory of Selection.—Darwin's Study of Domestic Animals and Cultivated Plants.—Andreas Wagner's notions as to the Special Creation of Cultivated Organisms for the good of Man.—The Tree of Knowledge in Paradise.—Comparison between Wild and Cultivated Organisms.—Darwin's Study of Domestic Pigeons.—Importance of Pigeon Breeding.— Common Descent of all Races of Pigeons.

DURING the thirty years, from 1830 until 1859, when Darwin's work appeared, the ideas of creation introduced by Cuvier remained predominant in the sciences of organic nature. People rested satisfied with the unscientific assumption, that in the course of the earth's history, a series of inexplicable revolutions had periodically annihilated the whole world of animals and plants, and that at the end of each revolution, and the beginning of a new period, a new enlarged, and improved edition of the organic population had appeared. Although the number of these editions of creation was altogether problematical, and in truth could not be fixed at all, and although the numerous advances which, during this time, were made in all the departments of zoology and botany demonstrated more and more that Cuvier's hypothesis was unfounded and untenable, and that Lamarck's natural theory of development was nearer the truth, yet the former maintained its authority almost universally among biologists. This must, above all, be ascribed to the veneration which Cuvier had acquired, and strikingly illustrates how injurious to the progress of humanity a faith in any definite authority may become. Authority, as Goethe once admirably said, perpetuates the individual, which as an individual should pass away, rejects and allows to pass that which should be held fast, and is the main obstacle to the advance of humanity.

It is only by having regard to the great weight of Cuvier's authority, and to the mighty potency of human indolence, which is with difficulty induced to depart from the broad and comfortable way of everyday conceptions, and to enter upon new paths not yet made easy, that we can comprehend how it is that Lamarck's Theory of Descent did not gain its due recognition until 1859, after Darwin had given it a new foundation. The soil had long been prepared for it by the works of Charles Lyell, another English naturalist, whose views are of great importance for the natural history of creation, and must accordingly here be briefly explained.

In 1830 Charles Lyell published, under the title of "Principles of Geology," a work in which he thoroughly reformed the science of Geology and the history of the earth's development, and effected this reform in a manner similar to that in which, thirty years later, Darwin in his work reformed the science of Biology. Lyell's great treatise, which radically destroyed Cuvier's hypothesis of creation, appeared in the same year in which Cuvier celebrated his triumph over the nature-philosophy, and established his supremacy in the domain of morphology for the following thirty years. Whilst Cuvier, by his artificial hypothesis of creation and his theory of catastrophes connected with it, directly obstructed the path of the theory of natural development, and cut off all chance of a natural explanation, Lyell once more opened a free road, and brought forward convincing geological evidence to show that Cuvier's dualistic conceptions were as unfounded as they were superfluous. He demonstrated that those changes of the earth's surface, which are still taking place before our eyes, are perfectly sufficient to explain everything we know of the development of the earth's crust in general, and that it is superfluous and useless to seek for mysterious causes in inexplicable revolutions. He showed that we need only have recourse to the hypothesis of exceedingly long periods of time in order to explain the formation of the crust of the earth in the simplest and most natural manner by means of the very same causes which are still active. Many geologists had previously imagined that the highest chains of mountains which rise on the surface of the earth could owe their origin only to enormous revolutions transforming a great part of the earth's surface, especially to colossal volcanic eruptions. Such chains of mountains as those of the Alps or the Cordilleras were believed to have arisen direct from the fiery fluid of the interior of the earth, through an enormous chasm in the broken crust. Lyell, on the other hand, showed that we can explain the formation of such enormous chains of mountains quite naturally by the same slow and imperceptible risings and depressions of the earth's surface which are still continually taking place, and the causes of which are by no means miraculous. Although these depressions and risings may perhaps amount only to a few inches, or at most a few feet, in the course of a century; still, in the course of some millions of years they are perfectly sufficient to raise up the highest chains of mountains, without the aid of mysterious and incomprehensible revolutions. In like manner, the meteorological action of the atmosphere, the influence of rain and snow, and, lastly, the breakers on the coasts, which by themselves seem to produce an insignificant effect, must cause the greatest changes if we only allow sufficiently long periods for their action. The multiplication of the smallest causes produces the greatest effects. Drops of water produce a cavity in a rock.

I shall afterwards be obliged again to recur to the immeasurable length of geological periods which are necessary for this purpose, for, as we shall see, Darwin's theory, as well as that of Lyell, renders the assumption of immense periods absolutely necessary. If the earth and its organisms have actually developed in a natural way, this slow and gradual development must certainly have taken a length of time which surpasses our powers of comprehension. But as many men see in this very circumstance one of the principal difficulties in the way of those theories of development, I beg leave here to remark that we have not a single rational ground for conceiving the time requisite to be limited in any way. Not only many ordinary persons, but even eminent naturalists, make it their chief objection to these theories, that they arbitrarily claim too great a length of time: yet the ground of objection is scarcely intelligible. For it is absolutely impossible to see what can, in any way, limit us in assuming long periods of time. We have long known, even from the structure of the stratified crust of the earth alone, that its origin and the formation of neptunic rocks from water must have taken, at least, several millions of years. From a strictly philosophical point of view, it makes no difference whether we hypothetically assume for this process ten millions or ten thousand billions of years. Before us and behind us lies eternity. If the assumption of such enormous periods is opposed to the feelings of many, I regard this simply as the consequence of false notions which are impressed upon us from our earliest youth concerning the short history of the earth, which is said to embrace only a few thousands of years. Albert Lange, in his "History of Materialism,"[12] has convincingly shown that from a strictly philosophical point of view it is far less objectionable in a scientific hypothesis to assume periods which are too long than periods which are too short. Every process of development is the more intelligible the longer it is assumed to last. A short and limited period is the most improbable.

I have no space here to enter minutely into Lyell's great work, and will therefore mention only its most important result, which is, that he completely refuted Cuvier's history of creation with its mythical revolutions, and established in its place the constant and slow transformation of the earth's crust by the continued action of forces, which are still working on the earth's surface, viz., the movement of water and the volcanic fluid of the interior of earth. Lyell thus demonstrated a continuous and uninterrupted connection of the whole history of the earth, and he proved it so irrefutably, and established so convincingly the supremacy of the "existing causes," that is, of the causes which are still active in the transformation of the earth's crust, that Geology in a short time completely renounced Cuvier's hypothesis.

Now, it is remarkable that Palaeontology, the science of petrifactions, so far as it was pursued by botanists and zoologists, remained apparently unaffected by this great progress in geology. Biology still continued to assume repeated new creations of the whole animal and vegetable kingdoms, at the beginning of every new period of the earth's history, although this hypothesis of individual creations, shoved into the world one after the other, without the assumption of Cuvier's cataclysms, became pure nonsense, and lost its foundation. It is evidently perfectly absurd to assume a distinct new creation of the whole world of animals and plants at definite epochs, without the crust of the earth itself experiencing any considerable general revolution. And although this conception is most closely connected with Cuvier's theory of catastrophes, still it prevailed when the latter had been completely destroyed and abandoned.

It was reserved for the great English naturalist, Charles Darwin, to remove this contradiction, and to show that the organic beings of the earth have a history as continuous and connected as the inorganic crust of the earth; that animals and plants have arisen from one another by as gradual a transmutation as that by which the varying forms of the earth's crust, the forms of the continents, and of the seas surrounding and separating them, have arisen out of earlier and quite different forms. In this respect we may truly say that in the domain of Zoology and Botany Darwin made the same progress as Lyell, his great countryman, in the domain of Geology. Both proved the *uninterrupted connection of the historical development*, and demonstrated a gradual transmutation of the different conditions succeeding one another.

The special merit of Darwin, as I have already remarked in a preceding chapter, is twofold. In the first place, he has treated the Theory of Descent, put forth by Lamarck and Goethe, in a much more comprehensive manner, as a whole, and carried it out in a much more connected manner, than had been done by any one of his predecessors. Secondly, he has established the causal foundation of this Theory of Descent by the Theory of Selection, which is peculiarly his own; that is, he has demonstrated the acting *causes of the changes* which the Theory of Descent simply stated, as *facts*. The Theory of Descent, introduced into Biology in 1809, by Lamarck, asserts that all the different species of animals and plants are descended from a single or some few most simple prototypes, produced by spontaneous generation. The Theory of Selection, established in 1859 by Darwin, shows us *why* this must be so; it points out the acting causes in a manner with which Kant would have been delighted, and indeed, in the domain of organic nature, Darwin has become the Newton whose advent Kant thought himself entitled prophetically to deny.

Now, before we approach Darwin's theory, it will perhaps be of interest to notice a few details as to the personal character of this great naturalist, as to his life, and the way in which he was led to form his doctrine. Charles Robert Darwin was born at Shrewsbury, on the Severn, on the 12th of February, 1809; therefore, at present he is sixty-three years old. In his seventeenth year (1825) he entered the University of Edinburgh, and two years later Christ's College, Cambridge. When scarcely twenty-two years old, in 1831, he was invited to take part in a scientific expedition which was sent out by England, in order to survey accurately the southernmost point of South America, and to examine several parts of the South Seas. This expedition, like many other voyages of inquiry fitted out in a praiseworthy manner by England, had scientific objects, and at the same time was intended to solve practical problems relating to navigation. The vessel, commanded by Captain Fitzroy, appropriately bore the symbolic name of the *Beagle*. The voyage of the *Beagle*, which lasted five years, was of the highest importance to the full development of Darwin's genius; for in the very first year, when he set his foot on the soil of South America, the outline of the doctrine of descent dawned upon him. Darwin himself has described this voyage in a work which is written in a very attractive style, and the perusal of which I strongly recommend to the reader. This book of travel, which lies far above the usual average in interest, not only shows in a very charming manner Darwin's amiable character, but we can in many ways recognize the various steps by which he arrived at his conceptions. The result of the voyage was, first, a large scientific work, the zoological and geological portion of which belong in a great measure to Darwin; and secondly, a celebrated work by him alone on Coral Reefs, which in itself would have sufficed to secure to him a lasting reputation. It is well known that the islands in the South Seas consist for the most part of coral reefs, and are surrounded by them. Formerly no satisfactory explanation could be given of their different and remarkable forms, and of their relation to those islands which are not formed of corals. It was reserved for Darwin to solve this difficult problem, for together with the constructive action of the coral zoophytes, he assumed geological risings and depressions of the bottom of the sea to account for the origin of the different forms of reefs. Darwin's Theory of the Origin of Coral Reefs, like his later one as to the Origin of Organic Species, is a theory which fully explains the phenomenon, and for this purpose assumes only the simplest natural causes, without hypothetically supporting it with any unknown processes. Among the remaining works of Darwin, I must not pass over his excellent monograph on Cirrhipedia, a curious class of marine animals, which in their outward appearance resemble mussels, and were actually considered by Cuvier as Molluscs possessing two shells, while in truth they belonged to the Crustacea (crabs).

The extraordinary hardships to which Darwin had been exposed during his voyage in the *Beagle* had injured his health to such a degree, that after his return home he was obliged to withdraw from the restless turmoil of London life, and since then has lived in quiet retirement on his estate at

Down, near Bromley, in Kent. This seclusion from the restless activity of the great city certainly exercised a beneficial influence upon Darwin, and it is probable that we owe to it, at least partially, the formation of the Theory of Selection. Undisturbed by the various engagements which in London would have wasted his strength, he was enabled to concentrate his attention upon the great problem to which his mind had been turned during his voyage in the *Beagle*. In order to show what kind of observations during the voyage principally gave rise to the fundamental idea of the Theory of Selection, and in what manner he afterwards worked it out, I shall insert here a passage from a letter which he addressed to me on the 8th of October, 1864.

Letter from Charles Darwin to Haeckel, 8th October, 1864.

"In South America three classes of facts were brought strongly before my mind. *Firstly*, the manner in which closely allied species replace species in going southward. *Secondly*, the close affinity of the species inhabiting the islands near South America to those proper to the continent. This struck me profoundly, especially the difference of the species in the adjoining islets in the Galopagos Archipelago. *Thirdly*, the relation of the living Edentata and Rodentia to the extinct species. I shall never forget my astonishment when I dug out a gigantic piece of armour like that of the living armadillo.

"Reflecting on these facts, and collecting analogous ones, it seemed to me probable that allied species were descended from a common parent. But for some years I could not conceive how each form became so excellently adapted to its habits of life. I then began systematically to study domestic productions, and after a time saw clearly that man's selective power was the most important agent. I was prepared, from having studied the habits of animals, to appreciate the struggle for existence, and my work in geology gave me some idea of the lapse of past time. Therefore, when I happened to read "Malthus on Population," the idea of natural selection flashed on me. Of all the minor points, the last which I appreciated was the importance and cause of the principle of divergence."

During the leisure and retirement in which Darwin lived after his return, he occupied himself, as we see from this letter, first and specially with the study of organisms in their cultivated state; that is, domestic animals and garden plants. This was undoubtedly the most likely way to arrive at the Theory of Selection. In this, as in all his labours, Darwin proceeded with extreme care and accuracy. With wonderful caution and self-denial, he published nothing on this subject during a period of twenty-one years, from 1837 to 1858, not even a preliminary sketch of his theory, which he had written as early as 1844. He was always anxious to collect still more certain experimental proofs, in order to be able to establish his theory in a complete form, and on the broadest possible foundation of experience. While he was thus aiming at the greatest possible perfection, which might perhaps have led him never to publish his theory at all, he was fortunately disturbed by a countryman of his, who, independently of Darwin, had discovered the

Theory of Selection, and in 1858 sent its outlines to Darwin himself, with the request to hand them to Lyell for publication in some English journal. This was Alfred Wallace, one of the boldest and most distinguished scientific travellers of modern times. For many years Wallace had wandered alone in the wilds of the Sunda Islands, dense primitive forests of the Indian Archipelago; and during close and comprehensive study of one of the richest and most interesting parts of the earth, with its great variety of animals and plants, he had arrived at exactly the same general views regarding the origin of organic species as Darwin. Lyell and Hooker, both of whom had long known Darwin's work, now induced him to publish a short extract from his manuscripts simultaneously with the manuscript sent him by Wallace. They appeared in the *Journal of the Linnean Society*, August, 1858.

Darwin's great work "On the Origin of Species," in which the Theory of Selection is carried out in detail, appeared in November, 1859. Darwin himself, however, characterizes this book (of which a fifth edition appeared in 1869, and the German translation by Bronn as early as 1860)[1] as only a preliminary extract from a larger and more detailed work, which is to contain a mass of facts in favour of his theory, and comprehensive and experimental proofs. The first part of the larger work promised by Darwin appeared in 1868, under the title, 'The Variations of Animals and Plants in the State of Domestication," and has been translated into German by Victor Carus.[14] It contains a rich abundance of the most valuable evidence as to the extraordinary changes of organic forms which man can produce by cultivation and artificial selection. However much we are indebted to Darwin for this abundance of convincing facts, still we do not by any means share the opinion of those naturalists who hold that the Theory of Selection requires for its actual proof these further details. It is our opinion that Darwin's first work, which appeared in 1859, already contains sufficient proof. The unassailable strength of his theory does not lie in the immense amount of individual facts that may be adduced as proofs, but in the harmonious connection of all the great and general phenomena of organic nature, which agree in bearing testimony to the truth of the Theory of Selection.

Darwin, at first, intentionally did not notice the important conclusion from his Theory of Descent, namely, the descent of the human race from other mammals. It was not till this highly important conclusion had been definitely established by other naturalists as the necessary sequel of the doctrine of descent, that Darwin himself expressly endorsed it, and thereby completed his system. This was done in the highly interesting work, "The Descent of Man, and Sexual Selection," which appeared as late as 1871, and has likewise been translated into German by Victor Carus.[48]

The careful study which Darwin devoted to *domestic animals and cultivated plants* was of the greatest importance in establishing the Theory of Selection. The infinitely varied changes of form which man has produced in these domesticated organisms by artificial selection are of the very highest importance for a right understanding of animal and vegetable forms; and yet this study has, down to the most recent times, been most grossly neglected by zoologists and botanists. Without entering upon the discussion of the significance to be attached to the idea of species itself, they have filled not only bulky volumes, but whole libraries, with descriptions of individual species, and with most childish controversies as to whether these species are good, or tolerably good, and bad, or tolerably bad. If naturalists instead of spending their time on these useless fancies had duly studied cultivated organisms, and had examined the transmutation of the living forms, instead of the individual dead ones, they would not have been led captive so long by the fetters of Cuvier's dogma. But as cultivated organisms are so extremely inconvenient to the dogmatic conception of the permanence of species, naturalists to a great extent intentionally did not concern themselves about them, and even celebrated naturalists have often expressed the opinion that cultivated organisms, domesticated animals and garden plants, are artificial productions of man, and that their formation and transformation could not decide anything about the nature of species and about the origin of the forms of species that live in a natural state.

This perverse view went so far that, for example, Andreas Wagner, a zoologist of Munich, quite seriously made the following ridiculous assertion:— "Animals and plants in their wild state have been called into being by the Creator as distinctly different and unchangeable species; but in the case of domestic animals and cultivated plants this was not necessary, because he formed them from the beginning for the use of man. The Creator formed man out of a clod of earth, breathed the living breath into his nostrils, and then created for him the different useful domestic animals and garden plants, among which he thought well to save himself the trouble of distinguishing species." Unfortunately, Andreas Wagner does not tell us whether the *Tree of Knowledge* in Paradise was a "good" wild species, or, as a cultivated plant, "no species" at all. As the Tree of Knowledge was placed by the Creator in the centre of Paradise, we might be inclined to believe that it was a highly favoured cultivated plant, and therefore no species at all. But since, on the other hand, the fruit of the Tree of Knowledge was forbidden to man, and since many men, as Wagner himself clearly shows, have never eaten of the fruit, it was evidently not created for the use of man, and therefore in all probability was a *real species*! What a pity Wagner has not given us any information about this important and difficult problem!

Now, however ridiculous this view may appear to us, it is only the logical sequence of a false view (which is widely spread) of the special nature of cultivated organisms, and one may occasionally hear similar objections from naturalists of great reputation. I must most decidedly, and at once, condemn this utterly false conception. It is the same perverseness which is committed by physicians who maintain that diseases are artificial productions, and not natural phenomena. It has been a work of hard labour to combat this prejudice, and it is only in recent times that men have generally adopted the view that diseases are nothing but natural changes of the organisms, or really natural phenomena of life, which are produced by changed and abnormal conditions of existence. Disease, therefore, is not a life beyond Nature's realm (vita praeter naturam), as the early physicians used to say, but a natural life under conditions which produce illness and threaten the body with danger. Just in the same manner, cultivated organic forms are not artificial works of man, but natural productions which have arisen under the influence of peculiar conditions of life. Man by his culture can never directly produce a new organic form, but he can breed organisms under new conditions of life, which are such as to influence and transform them. All domestic animals and all garden plants are originally descended from wild species, which have been transformed by the peculiar conditions of culture.

A thorough comparison of cultivated forms (races and varieties) with organisms not altered by cultivation (species and varieties), is of the utmost importance to the theory of selection. What is most surprising in such a comparison is the remarkably short time in which man can produce a new form, and the high degree in which this form, produced by man, can deviate from the original form. While wild animals and plants, one year after another, appear to the zoologist and botanist approximately in the same form, so as to have given rise to the false doctrine of the constancy of species, domestic animals and garden plants, on the other hand, display the greatest changes within a few years. The perfection which gardeners and farmers have attained in the art of selection now enables them, in the space of a few years, arbitrarily to create entirely new animal and vegetable forms. For this purpose it is only necessary to keep and propagate the organism under the influence of special conditions—which are capable of producing new formations—and even at the end of a few generations new species may be obtained, which differ from the original form in a much higher degree than so-called good species in a wild state differ from one another. This fact is extremely important, and we cannot lay sufficient stress upon it. The assertion is not true that cultivated forms descended from one and the same primary form do not differ from one another as much as wild animal and vegetable species differ among themselves. If we only make comparisons, without prejudice, we can very easily perceive that a number of races or varieties which have been derived from a single cultivated form, within a short series of years, differ from one another in a higher degree than so-called good species (bonae species), or even different genera of one family, in the wild state.

In order to establish this extremely important fact as firmly as possible by experiments, Darwin decided to make a special study of the whole extent of variation in form in a single group of domesticated animals, and for this purpose he chose the *domestic pigeons*, which are in many respects especially suited for such a study. For a long time he kept on his estate all possible races and varieties of pigeons which he was able to procure, and he was helped in this by rich contributions from all parts of the world. He also joined two London pigeon clubs, the members of which passionately, and with truly artistic skill, carry on the breeding of the different forms of pigeons. Lastly, he formed connections with some of the most celebrated pigeon-fanciers; so that he could command the richest experimental material.

The art of, and fancy for, pigeon breeding is very ancient. Even more than 3,000 years before Christ, it was carried on by the Egyptians. The Romans, under the emperors, laid out enormous sums upon the breeding of pigeons, and kept accurate pedigrees of their descent, just as the Arabs keep genealogical pedigrees of their horses, and the Mecklenburg aristocracy of their own ancestors. In Asia, too, among the wealthy princes, pigeon breeding was a very ancient fancy; in 1600, the court of Akber Khan possessed more than 20,000 pigeons. Thus in the course of several centuries, and in consequence of the various methods of breeding practised in the different parts of the world, there has arisen out of one single originally tamed form, an immense number of different races and varieties, which in their most divergent forms are extremely different from one another, and are often curiously characterized.

One of the most striking races of pigeons is the well-known fan-tailed pigeon, which spreads its tail like the peacock, and carries a number of (from thirty to forty) feathers placed in the form of radii, while other pigeons possess much fewer tail feathers—generally twelve. We may here mention that the number of feathers on the tails of birds is considered by naturalists of great value as a systematic distinction, so that whole orders can thereby be distinguished. For example, singing birds, almost without exception, possess twelve tail feathers; chirping birds (Strisores) ten, etc. Several races of pigeons, moreover, are characterized by a tuft of neck feathers, which form a kind of periwig; others by grotesque transformation of their beaks and feet, by peculiar and often very remarkable decorations, as, for example, skinny lappets, which develop on the head; by a large crop, which is formed by the gullet being strongly inclined forward, etc. Remarkable, also, are the strange habits which many pigeons have acquired; for example, the turtle pigeons and the trumpeters with their musical accomplishments, the carriers with their topographical instinct. The tumblers have the strange habit of ascending into the air in great numbers, then turning over and falling down through the air as if dead. The ways and habits of these endless races of pigeons—the form, size, and colour of the individual parts of their bodies, and their proportions, differ in a most astonishing degree from one another; in a much higher degree than is the case with the so-called good species, or even with the perfectly distinct genera, of wild pigeons. And what is of the greatest importance, is the fact that these differences are not confined to the external form, but extend even to the most important internal parts; there even occur great modifications of the skeleton and of the muscular tissues. For example, we find great differences in the number of vertebrae and ribs, in the size and shape of the gaps in the breast-bones, in the size and shape of the merry-thought, in the lower jaw, in the facial bones, etc. In short, the bony skeleton, which morphologists consider a very permanent part of the body, and which never varies to such an extent as the external parts—shows such great changes, that many races of pigeons might be described as special genera, and this would doubtless be done if all these different forms had been found in a wild and natural state.

How far the differences of the races of pigeons have been carried is best shown by the fact that all pigeon breeders are unanimously of opinion that each peculiar or specially marked race of pigeons must be derived from a corresponding wild original species. It is true every one assumes a different number of original species. Yet Darwin has most convincingly and acutely proved that all these pigeons, without exception, must be derived from a single wild primary species—from the blue rock-pigeon (*Columba livia*). In like manner, it can be proved of most of the domestic animals and cultivated plants, that all the different races are descendants of a single original wild species which has been brought by man into a cultivated condition.

An example similar to that of the domestic pigeons is furnished among mammals by our tame *rabbit*. All zoologists, without exception, have long considered it proved that all its races and varieties are descended from the common wild rabbit, that is, from a single primary species. And yet the extreme forms of these races differ to such a degree from one another, that every zoologist, if he met with them in a wild state, would unhesitatingly designate them not only as an entirely distinct "good species," but even as species of entirely different genera of the Leporid family. Not only does the colour, length of hair, and other qualities of the fur of the different tame races of rabbits vary exceedingly, and form extremely broad contrasts, but, what is still more important, the typical form of the skeleton and its individual parts do so also, especially the form of the skull and the jaw (which is of such importance in systematic arrangement); further, the relative proportion of the length of the ears, legs, etc. In all these respects the races of tame rabbits avowedly differ from one another far more than all the different forms of wild rabbits and hares which are scattered over all the earth, and are the recognized "good species" of the genus *Lepus*. And yet, in the face of these clear facts, the opponents of the theory of development maintain that the wild species are not descended from a common prototype, although they at once admit it in the case of the tame races. With opponents who so intentionally close their eyes against the clear light of truth, no further dispute can be carried on.

While in this manner it appears certain that the domestic races of pigeons, of tame rabbits, of horses, etc., notwithstanding the remarkable difference of their varieties, are descended in each case from but one wild, so-called "species"; yet, on the other hand, it is certainly probable that the great variety of races of some of the domestic animals, especially dogs, pigs, and oxen, must be ascribed to the existence of several wild prototypes, which have become mixed. It is, however, to be observed that the number of these originally wild primary species is always much smaller than that of the cultivated forms proceeding from their mingling and selection, and naturally they were originally derived from a single primary ancestor, common to the whole genus. In no case is each separate cultivated race descended from a distinct wild species.

In opposition to this, almost all farmers and gardeners maintain, with the greatest confidence, that each separate race bred by them must be descended from a separate wild primary species, because they clearly perceive the differences of the races, and attach very high importance to the inheritance of their qualities; but they do not take into consideration the fact that these qualities have arisen only by the slow accumulation of small and scarcely observable changes. In this respect it is extremely instructive to compare cultivated races with wild species.

Many naturalists, and especially the opponents of the Theory of Development, have taken the greatest trouble to discover some morphological or physiological mark, some characteristic property, whereby the artificially bred and cultivated races may be clearly and thoroughly distinguished from wild species which have arisen naturally. All these attempts have completely failed, and have led only with increasing certainty to the result, that such a distinction is altogether impossible. I have minutely discussed this fact, and illustrated it by examples in my criticism of the idea of species. (Gen. Morph. ii. 323–364.)

I may here briefly touch on yet another side of this question, because not only the opponents, but even a few of the most distinguished followers of Darwin—for example, Huxley—have regarded the phenomena of *bastard-breeding*, or *hybridism*, as one of the weakest points of Darwinism. Between cultivated races and wild species, they say, there exists this difference, that the former are capable of producing fruitful bastards, but that the latter are not. Two different cultivated races, or wild *varieties of one species*, are said in all cases to possess the power of producing bastards which can fruitfully mix with one another, or with one of their parent forms, and thus propagate themselves; on the other hand, *two really different species*, two cultivated or wild *species* of one genus, are said *never* to be able to produce from one another bastards which can be fruitfully crossed with one another, or with one of their parent species.

As regards the first of these assertions, it is simply refuted by the fact that there are organisms which do not mix at all with their own ancestors, and therefore can produce no fruitful descendants. Thus, for example, our cultivated guinea-pig does not bear with its wild Brazilian ancestor; and again, the domestic cat of Paraguay, which is descended from our European domestic cat, no longer bears with the latter. Between different races of our domestic dogs, for example, between the large Newfoundland dogs and the dwarfed lap-dogs, breeding is impossible, even for simple mechanical reasons. A particularly interesting instance is afforded by the Porto-Santo rabbit (*Lepus huxleyi*). In the year 1419, a few rabbits, born on board ship of a tame Spanish rabbit, were put on the island of Porto Santo, near Madeira. These little animals, there being no beasts of prey, in a short time increased so enormously that they became a pest to the country, and even compelled a colony to remove from the island. They still inhabit the island in great numbers; but in the course of four hundred and fifty years they have developed into a quite peculiar variety—or if you will have it, into a "good species"— which is distinguished by a peculiar colour, a rat-like shape, small size, nocturnal life, and extraordinary wildness. The most important fact, however, is that this new species, which I call *Lepus huxleyi*, no longer pairs with its European parent rabbit, and no longer produces bastards with it.

On the other hand, we now know of numerous examples of fruitful genuine bastards; that is, of mixings that have proceeded from the crossing of two entirely different species, and yet propagate themselves with one another as well as with one of their parent species. A number of such bastard species (species *Hybridae*) have long been known to botanists; for example, among the genera of the thistle (*Cirsium*), the laburnum (*Cytisus*), the bramble (*Rubus*), etc. Among animals also they are by no means rare, perhaps even very frequent. We know of fruitful bastards which have arisen from the crossing of two different species of a genus, as among several genera of butterflies (*Zygaena, Saturnia*), the family of carps, finches, poultry, dogs, cats, etc. One of the most interesting is the hare-rabbit (*Lepus darwinii*), the bastard of our indigenous hare and rabbit, many generations of which have been bred in France, since 1850, for gastronomic purposes. I myself possess such hybrids, the products of pure in-breeding, that is, both parents of which are themselves hybrids by a hare-father and a rabbit-mother. I possess them through the kindness of Professor Conrad, who has repeatedly made these experiments in breeding on his estate. The half-blood hybrid thus bred, which I name in honour of Darwin, appears to propagate itself through many generations by pure in-breeding, just as well as any genuine species. Although on the whole it is more like its mother (rabbit), still in the formation of the ears and of the hind-legs, it possesses distinct qualities of its father (hare). Its flesh has an excellent taste, rather resembling that of a hare, though the colour is more like that of a rabbit. But the hare (*Lepus timidus*) and the rabbit (*Lepus cuniculus*) are two species of the genus *Lepus*, so different that no systematic zoologist will recognize them as varieties of one species. Both species, moreover, live in such different ways, and in their wild state entertain so great an aversion towards one another, that they do not pair so long as they are left free. If, however, the newly-born young ones of both species are brought up together, this aversion is not developed; they pair with one another and produce the *Lepus darwinii*.

Another remarkable instance of the crossing of different species (where the two species belong even to different genera!) is furnished by the fruitful hybrids of sheep and goats which have for a long time been bred in Chili for industrial purposes. On what unessential circumstances in the sexual mingling the fertility of the different species depend, is shown by the fact that he-goats and sheep in their mingling produce fruitful hybrids, while the ram and she-goat pair very rarely, and then without result. The phenomena of hybridism to which undue importance has been erroneously attributed are thus utterly unmeaning, so far as the idea of species is concerned. The breeding of hybrids does not enable us, any more than other phenomena, thoroughly to distinguish cultivated races from wild species; and this circumstance is of the greatest importance in the Theory of Selection.

CHAPTER VII.

THE THEORY OF SELECTION (DARWINISM).

Darwinism (Theory of Selection) and Lamarckism (Theory of Descent).—The Process of Artificial Breeding.—Selection of the Different Individuals for After-breeding.—The Active Causes of Transmutation.—Change connected with Food, and Transmission by Inheritance connected with Propagation.—Mechanical Nature of these Two Physiological Functions.—The Process of Natural Breeding: Selection in the Struggle for Existence.—Malthus' Theory of Population.—The Proportion between the Numbers of Potential and Actual Individuals of every Species of Organisms.—General Struggle for Existence, or Competition to attain the Necessaries of Life.—Transforming Force of the Struggle for Existence.—Comparison of Natural and Artificial Breeding.—Selection in the Life of Man.—Military and Medical Selection.

It is, properly speaking, not quite correctly that the Theory of Development, with which we are occupied in these pages, is usually called Darwinism. For, as we have seen from the historical sketch in the previous chapters, the most important foundation of the Theory of Development—that is, the Doctrine of Filiation, or Descent—had already been distinctly enunciated at the beginning of our century, and had been definitely introduced into science by Lamarck. The portion of the Theory of Development which maintains the common descent of all species of animals and plants from the simplest common original forms might, therefore, in honour of its eminent founder, and with full justice, be called *Lamarckism*, if the merit of having carried out such a principle is to be linked to the name of a single distinguished naturalist. On the other hand, the Theory of Selection, or breeding, might be justly called *Darwinism*, being that portion of the Theory of Development which shows us in what way and *why* the different species of organisms have developed from those simplest primary forms. (Gen. Morph. ii. 166.)

It is true we find the first trace of an idea of natural selection even forty years before the appearance of Darwin's work. For in the year 1818 there was published a paper "On a woman of the white race whose skin partly resembled that of a negro," which had been read before the Royal Society as early as 1813. Its author, Dr. W. C. Wells, states that negroes and mulattoes are distinguished from the white race by their immunity from certain tropical diseases. On this occasion he remarks that all animals have a tendency to change up to a certain degree, and that farmers, by availing themselves of this tendency, and also by selection, improve their domestic animals; and then he adds, that what is done in this latter case "by art, seems to be done with equal efficiency, though more slowly, by nature, in the formation of varieties of mankind fitted for the country which they inhabit. Of the accidental varieties of man which would occur among the first few and scattered inhabitants of the middle regions of Africa, some one would be better fitted than the others to bear the diseases of the country. This race would consequently multiply, while the others would decrease; not only from their inability to sustain the attacks of disease, but from their incapacity of contending with their more vigorous neighbours. The colour of this vigorous race I take for granted, from what has been already said, would be dark. But the same disposition to form varieties still existing, a darker and a darker race would in the course of time occur; and as the darkest would be the best fitted for the climate, this would at length become the most prevalent, if not the only race, in the particular country in which it had originated." He then extends these same views to the white inhabitants of colder climates. Although Wells clearly expresses and recognizes the principle of natural selection, yet it is applied by him only to the very limited problem of the origin of human races, and not at all to that of the origin of animal and vegetable species. Darwin's great merit in having independently developed the Theory of Selection, and having brought it to complete and well merited recognition, is as little affected by the earlier and long forgotten remark of Wells, as by some other fragmentary observations about natural selection made by Patrick Mathew, and hidden in his book on "Timber for Shipbuilding, and the Cultivation of Trees," which appeared in 1831. The celebrated traveller, Alfred Wallace, who developed the Theory of Selection independently of Darwin, and had published it in 1858, simultaneously with Darwin's first contribution, likewise stands far behind his greater and elder countryman in regard to profound conception, as well as to extended application of the theory. In fact Darwin, by his extremely comprehensive and ingenious development of the whole doctrine, has acquired a fair claim to see the theory connected with his own name.

This Theory of Selection, Darwinism in its proper sense, to the consideration of which we now turn our attention, rests essentially (as has already been intimated in the last chapter) upon the comparison of those means which man employs in the breeding of domestic animals and the cultivation of garden plants, with those processes which in free nature, outside the cultivated state, lead to the coming into existence of new species and new genera. We must therefore, in order to understand the latter processes, first turn to the artificial breeding by man, as was, in fact, done by Darwin himself. We must inquire into the results to which man attains by his artificial breeding, and what means are applied in order to obtain those results; and we must then ask ourselves, "Are there in nature similar forces and causes acting similarly to those resorted to by man?"

First, in regard to artificial breeding, we start from the fact last discussed above, viz., that its products in some cases differ from one another much more than the productions of natural breeding. It is a fact that races or varieties often differ from one another in a much greater degree and in much more important qualities than many so-called species, or "good species,"—nay, sometimes even more than so-called "good genera" in their natural state. Compare, for example, the different kinds of apples which the art of horticulture has derived from one and the same original apple-form, or compare the different races of horses which their breeders have derived from one and the same original form of horse, and it will be easily observed that the differences of the most different forms are extremely important, and much more important than the so-called "specific differences," which are referred to by zoologists and botanists when comparing wild forms for the purpose of distinguishing several so-called "good species."

Now, by what means does man produce this extraordinary difference or divergence of several forms which are proved to be descended from the same primary form? In order to answer this question, let us follow a gardener who desires to produce a new form of a plant, which is distinguished by the beautiful colour of its flowers. He will first of all make a selection from a great number of plants which are seedlings from one and the same parent. He will pick out those plants which exhibit most distinctly the colour of flower he desires. The colour of flowers is a very changeable thing. Plants, for example, which as a rule have a white flower, frequently show deviations into the blue or red. Now, supposing the gardener wishes to obtain the red colour in a plant usually producing white flowers, he will very carefully, from among the many different individuals which are the descendants of one and the same seed-plant, select those which most distinctly show a reddish tint, and sow them exclusively, in order to produce new individuals of the same kind. He would cast aside and no longer cultivate the other seedlings which show a white or less distinct red colour. He will propagate exclusively the individual plants whose blossoms show the red most markedly, and he will sow the seeds produced by these selected plants. From the seedlings of this second generation, he will again carefully select those in which the red, which is now visible in the majority of them, is most distinctly displayed. If such a selection is carried on during a series of six or ten generations, and if the flower which shows the deepest red is most carefully selected, the gardener in the sixth or tenth generation will obtain the desired plants with flowers of a pure red.

The farmer wishing to breed a special race of animals, for example, a kind of sheep distinguished by particularly fine wool, proceeds in the same manner. The only process applied in the improvement of wool consists in this, that the farmer with the greatest care and perseverance selects from a whole flock of sheep those individuals which have the finest wool. These only are used in breeding, and among the descendants of these selected sheep, those again are chosen which have the finest wool, etc. If this careful selection is carried on through a series of generations, the selected breeding-sheep are in the end distinguished by a wool which differs very strikingly from the wool of the original parent, and this is exactly the advantage which the breeder desired.

The differences of the individuals that come into consideration in this artificial selection are very slight. An ordinary unpractised man is unable to discover the exceedingly minute differences of individuals which a practised breeder perceives at the first glance. The business of a breeder is not easy; it requires an exceedingly sharp eye, great patience, and an extremely careful manner of treating the organisms to be bred. In each individual generation, the differences of individuals are perhaps not seen at all by the uninitiated; but by the accumulation of these minute differences during a series of generations, the deviation from the original form becomes in the end very great. It becomes so great that the artificially produced form may in the end differ far more from the original form than do two so-called "good species" in their natural state. The art of breeding has now made such progress, that man can often at discretion produce certain peculiarities in cultivated species of animals and plants. To practised gardeners and farmers, you may give distinct commissions, and say, for example, I wish to have this species of plant with this or that colour, and with this or that shape. Where breeding has reached the perfection which it has attained in England, gardeners and farmers are frequently able to furnish to order the desired result within a definite period, that is, at the end of a number of generations. Sir John Sebright, one of the most experienced English pigeon-breeders, could assert that in three years he would produce any form of feather, but that he required six years to obtain any desired form of the head and beak. In the process of breeding the merino-sheep of Saxony, the animals are three times placed on a table beside one another, and most carefully compared and studied.

Each time only the best sheep with the finest wool are selected, so that in the end, out of a great multitude, there remain only some few animals, but their wool is exquisitely fine, and only these last are used in breeding. We see, therefore, that the causes through which, in artificial breeding, great effects are produced, are unusually simple, and these great effects are obtained simply by accumulating the differences which in themselves are very insignificant, and become surprisingly increased by a continually repeated selection.

Before we pass on to a comparison of this artificial with natural breeding, let us see what natural qualities of the organisms are made use of by the artificial breeder or cultivator. We can trace all the different qualities which here come into play to physiological fundamental qualities of the organism, which are common to all animals and plants, and are most closely connected with the functions of *propagation* and *nutrition*. These two fundamental qualities are *transmissivity*, or the capability of *transmitting by inheritance*, and *mutability*, or the capability of *adaptation*. The breeder starts from the fact that all the individuals of one and the same species are different, though in a very slight degree, a fact which is as true of organisms in a wild as in a cultivated state. If you look about you in a forest consisting of only a single species of tree, for example of beech, you will certainly not find in the whole forest two trees of this kind which are absolutely identical or perfectly equal in the form of their branches, the number of their branches and leaves, blossoms and fruits. Special differences occur everywhere, just as in the case of men. There are no two men who are absolutely identical, perfectly equal in size, in the formation of their faces, the number of their hairs, their temperament, character, etc. The very same is true of individuals of all the different species of animals and plants. It is true that in most organisms the differences are very trifling to the eye of the uninitiated. Everything here essentially depends on the exercise of the faculty of discovering these often very minute differences of form. The shepherd, for example, knows every individual of his flock, solely by accurately observing their features, while the uninitiated are incapable of distinguishing at all the different individuals of one and the same flock. This fact of the individual difference is the extremely important foundation on which the whole of man's power of breeding rests. If individual differences did not exist everywhere, man would not be able to produce a number of different varieties or races from one and the same original stock. We must, at the outset, hold fast the principle that the phenomenon is quite universal; we must necessarily assume it even where, with the imperfect capabilities of our senses, we are unable to discover differences. Among the higher plants (the phanerogams, or flower-plants), where the individual stocks show such numerous differences in the number of branches or leaves, and in the formation of the stem and branches, we can almost always easily perceive these differences. But this is not the case in the lower plants, such as mosses, algae, fungi, and in most animals, especially the lower ones. The distinction of all the individuals of one species is here, for the most part, extremely difficult or altogether impossible. But there is no reason for ascribing individual differences only to those organisms in which we can perceive them at once. We may, on the contrary, with full certainty assume such individuality as a universal quality of all organisms, and we can do this all the more surely since we are able to trace the mutability of individuals to the mechanical conditions of nutrition. We can show that by influencing nutrition we are able to produce striking individual differences where they would not exist if the conditions of nutrition had not been altered. The many complicated conditions of nutrition are never absolutely identical in two individuals of a species.

Now, just as we see that the mutability or capability of adaptation has a causal connection with the general relations of nutrition in animals and plants, so too we find the second fundamental phenomenon of life, with which we are here concerned, namely, the capability of *transmitting by inheritance*, to have a direct connection with the phenomenon of *propagation*. The second thing that a farmer or gardener does in artificial breeding, after he has selected, and has consequently availed himself of the mutability, is to endeavour to hold fast and develop the modified forms by Inheritance. He starts from the universal fact that children resemble their parents, that "the apple does not fall far from the tree." This phenomenon of Inheritance has hitherto been scientifically examined only to a very small extent, which may partly arise from the fact that the phenomenon is of such everyday occurrence. Every one considers it quite natural that every species should produce its like; that a horse should not suddenly produce a goose, or a goose a frog. We are accustomed to look upon these everyday occurrences of Inheritance as self-evident. But this phenomenon is not so simply self-evident as it appears at first sight, and in the examination of Inheritance the fact is very frequently overlooked that the different descendants, derived from one and the same parents, are in reality *never* quite identical, and also never absolutely like the parents, but are always slightly different. We cannot formulate the principle of Inheritance, as "Like produces like," but we must limit the expression to "Similar things produce similar things." The gardener, as well as the farmer, avails himself of the fact of Inheritance in its widest form, and indeed with special regard to the fact that not only those qualities of organisms are transmitted by inheritance which they have inherited from *their* parents but those also which they themselves have *acquired*. This is an important point upon which very much depends. An organism can transmit to its descendants not only those qualities of form, colour, and size which it has inherited from its parents, but it can also transmit changes of these qualities, which it has acquired during its own life through the influence of outward circumstances, such as climate, nourishment, training, etc.

These are the two fundamental qualities of animals and plants of which the breeder must avail himself in order to produce new forms. The theoretical principle of breeding is, indeed, extremely simple, but in detail the practical application of this simple principle is difficult and immensely complicated. A thoughtful breeder, acting according to a definite plan, must understand the art of correctly estimating, in every case, the general interaction between the two fundamental qualities of heirship and mutability.

Now, if we examine the real nature of those two important properties of life, we find that we can trace them, like all physiological functions, to physical and chemical causes, to the properties and the phenomena of motion of those substances of which the bodies of animals and plants consist. As we shall hereafter have to show in the more accurate consideration of these two functions, the transmission by *Inheritance*, if we express ourselves quite generally, is essentially dependent upon the material continuity and partial identity of the matter in the producing and produced organism, the parents and the child. In every act of breeding a certain quantity of protoplasm or albuminous matter is transferred from the parents to the child, and along with it there is transferred the individually *peculiar molecular motion.* These molecular phenomena of motion in the protoplasm, which call forth the phenomena of life, and are their active and true cause, differ more or less in all living individuals; they are of infinite variety.

Adaptation, or transmutation is, on the other hand, essentially the consequence of material influences, which the substance of the organism experiences from the material surrounding it,—in the widest sense of the word from the *conditions* of life. The external influences of the latter are communicated to the individual parts of the body by the molecular processes of nutrition. In every act of Adaptation the individual molecular motion of the protoplasm, peculiar to each part, disturbs and modifies the whole individual, or part of it, by mechanical, physical, or chemical influences. The innate, inherited vital actions of the protoplasm—that is, the molecular phenomena of motion of the smallest albuminous particles—are therefore more or less modified by it. The phenomenon of Adaptation, or transmutation, depends therefore upon the material influence which the organism experiences from its surroundings, or its conditions of existence; while the transmission by Inheritance is due to the partial identity of the producing and produced organism. These are the real, simple, mechanical foundations of the artificial process of breeding.

Now Darwin asked himself, Does there exist a similar process of selection in nature, and are there forces in nature which take the place of man's activity in artificial selection? Is there a natural tendency among wild animals and plants which acts selectingly, in a similar manner to the artificial selection practised by the designing will of man? All here depended upon the discovery of such a relation, and Darwin succeeded in this so satisfactorily, that we consider his theory of selection completely sufficient to explain, mechanically, the origin of the wild species of animals and plants. That relation which in free nature influences the forms of animals and plants, by selecting and transforming them, is called by Darwin the "*Struggle for Existence.*"

The "Struggle for Existence" has rapidly become a watchword of the day. Yet this designation is, perhaps, in many respects not very happily chosen, and the phenomena might probably have been more accurately described as "*Competition for the Means of Subsistence.*" For under the name of "Struggle for Life," many relations are comprehended which properly and strictly speaking do not belong to it. As we have seen from the letter inserted in the last chapter, Darwin arrived at the idea of the "Struggle for Existence" from the study of Malthus' book "On the Conditions and the Consequences of the Increase of Population." It was proved in that important work, that the number of human beings, on the average, increases in a geometrical progression, while the amount of articles of food increase only in an arithmetical progression.

This disproportion gives rise to a number of inconveniences in the human community, which cause among men a continual competition to obtain the necessary means of life, which do not suffice for all.

Darwin's theory of the struggle for life is, to a certain extent, a general application of Malthus' theory of population to the whole of organic nature. It starts from the consideration that the number of *possible* organic individuals which might arise from the germs produced, is far greater than the number of *actual* individuals which, in fact, do simultaneously live on the earth's surface. The number of possible or *potential individuals* is given us by the number of the eggs and organic germs produced by organisms. The number of these germs, from each of which, under favourable circumstances, an individual might arise, is very much larger than the number of real or actual *individuals*—that is, of those that really arise from these germs, come into life, and propagate themselves. By far the greater number of germs perish in the earliest stage of life, and it is only some favoured organisms which manage to develop, and actually survive the first period of early youth, and finally succeed in propagating themselves. This important fact is easily proved by a comparison of the number of eggs in a given species with the number of individuals which exist of this species. These numerical relations show the most striking contrast. There are, for example, species of fowls which lay great numbers of eggs, and yet are among the rarest of birds; and the bird which is said to be the commonest (the most widely spread) of all, the stormy petrel (*Procellaria glacialis*), lays only a single egg. The relation is the same in other animals. There are many very rare invertebrate animals, which lay immense quantities of eggs; and others again which produce only very few eggs, and yet are among the commonest of animals. Take, for example, the proportion which is observed among the human tape-worms. Each tape-worm produces within a short period millions of eggs, while man, in whom these tape-worms are lodged, forms a far smaller number of eggs, and yet fortunately there are fewer tape-worms than human beings. In like manner, among plants there are many splendid orchids, which produce thousands of seeds and yet are very rare, and some kinds of asters (Compositae), which have but few seeds, are exceedingly common.

This important fact might be illustrated by an immense number of examples. It is evidently, therefore, not the number of actually existing germs which indicates the number of individuals which afterwards come into life and maintain themselves in life; but rather the case is this, that the number of adult individuals is limited by other circumstances, especially by the relations in which the organism stands to its organic and inorganic surroundings. Every organism, from the commencement of its existence, struggles with a number of hostile influences: it struggles against animals which feed on it, and to which it is the natural food, against animals of prey and parasites; it struggles against inorganic influences of the most varied kinds, against temperature, weather, and other circumstances; but it also struggles (and this is much the most important!), above all, against organisms most like and akin to itself. Every individual, of every animal and vegetable species, is engaged in the fiercest competition with every other individual of the same species which lives in the same place with it. In the economy of nature the means of subsistence are nowhere scattered in abundance, but are very limited, and far from sufficient for the number of organisms which might develop from the germs produced. Therefore the young individuals of most species of animals and vegetables must have hard work in obtaining the means of subsistence; this necessarily causes a competition among them in order to obtain the indispensable supplies of life.

This great competition for the necessaries of life goes on everywhere and at all times, among human beings and animals as well as among plants; in the case of the latter this circumstance, at first sight, is not so clearly apparent. If we examine a field which is richly sown with wheat, we can see that of the numerous young plants (perhaps some thousands) which shoot up on a limited space, only a very small proportion preserve themselves in life. A competition takes place for the space of ground which each plant requires for fixing its root, a competition for sunlight and moisture. And in the same manner we find that, among all animal species, all the individuals of one and the same species compete with one another to obtain these indispensable means of life, or the conditions of existence in the wide sense of the word. They are equally indispensable to all, but really fall to the lot of only a few—"Many are called, but few are chosen." The fact of the great competition is quite universal. You need only to cast a glance at human society, where this competition exists everywhere, and in all the different branches of human activity. Here, too, a struggle is brought about by the free competition of the different labourers of one and the same class. Here too, as everywhere, this competition benefits the thing, or the work, which is the object of competition. The greater and more general the competition, the more quickly improvements and inventions are made in the branch of labour, and the higher is the grade of perfection of the labourers themselves.

The position of the different individuals in this struggle for life is evidently very unequal. Starting from the inequality of individuals, which is a recognized fact, we must in all cases necessarily suppose that all the individuals of one and the same species do not have equally favourable prospects. Even at the beginning they are differently placed in this competition by their different strengths and abilities, independently of the fact that the conditions of existence are different, and act differently at every point of the earth's surface. We evidently have an infinite combination of influences, which, together with the original inequality of the individuals during the competition for the conditions of existence, favour some individuals and prejudice others. The favoured individuals will gain the victory over the others, and while the latter perish more or less early, without leaving any descendants, the former alone will be able to survive and finally to propagate the species. As, therefore, it is clear that in the struggle for life the favoured individuals succeed in propagating themselves, we shall (even as the result of this relation) perceive in the next generation differences from the preceding one. Some individuals of this second generation, though perhaps not all of them, will, by inheritance, receive the individual advantage by which their parents gained the victory over their rivals.

But now—and this is a very important law of inheritance—if such a transmission of a favourable character is continued through a series of generations, it is not simply transmitted in the original manner, but it is constantly increased and strengthened, and in a last generation it attains a strength which distinguishes this generation very essentially from the original parent. Let us, for example, examine a number of plants of one and the same species which grow together in a very dry soil. As the hairs on the leaves of plants are very useful for receiving moisture from the air, and as the hairs on the leaves are very changeable, the individuals possessing the thickest hair on their leaves will have an advantage in this unfavourable locality where the plants have directly to struggle with the want of water, and in addition to this have to compete with one another for the possession of what little water there may be. These alone hold out, while the others possessing less hairy leaves perish; the more hairy ones will be propagated, and their descendants will, on the average, be more distinguished by their thick and strong hairs than the individuals of the first generation. If this process is continued for several generations in one and the same locality, there will arise at last such an increase of this characteristic, such an increase of the hairs on the surface of the leaf, that an entirely new species seems to present itself. It must here be observed, that in consequence of the interactions of all the parts of every organism, generally one individual part cannot be changed without at the same time producing changes in other parts. If, for instance, in our imaginary example, the number of the hairs on the leaves is greatly increased, a certain amount of nourishment is thereby withdrawn from other parts; the material which might be employed to form flowers or seeds is diminished, and a smaller size of the flower or seed will then be the direct or indirect consequence of the struggle for life, which in the first place only produced a change in the leaves. Thus the struggle for life, in this instance, acts as a means of selecting and transforming. The struggle of the different individuals to obtain the necessary conditions of existence, or, taking it in its widest sense, the inter-relations of organisms to the whole of their surroundings, produce mutations of form such as are produced in the cultivated state by the action of man's selection.

This agency will perhaps appear at first sight small and insignificant, and the reader will not be inclined to concede to the action of such relations the weight which it in reality possesses. I must therefore find space in a subsequent chapter to put forward further examples of the immense and far-reaching power of transformation exhibited in natural selection. For the present I will confine myself to simply placing side by side the two processes of artificial and natural selection, and clearly explaining the agreement and the differences of the two.

Both natural and artificial selection are quite simple natural, mechanical relations of life, which depend upon the *interaction* of two physiological functions, namely, on *Adaptation* and *Inheritance*, functions which, as such, must again be traced to physical and chemical properties of organic matter. The difference between the two forms of selection consists in this: in artificial selection the will of man makes the selection according to a *plan*, whereas in natural selection, the struggle for life (that universal inter-relation of organisms) acts *without a plan*, but otherwise produces quite the same result, namely, a selection of a particular kind of individuals for propagation. The alterations produced by artificial selection are turned to the advantage of *those who make the selection*; in natural selection, on the other hand, to the advantage of the *selected organism*.

These are the most essential differences and agreements of the two modes of selection; it must, however, be further observed that there is another difference, viz., in the duration of time required for the two processes of selection. Man in his artificial selection can produce very important changes in a very short time, while in natural selection similar results are obtained only after a much longer time. This arises from the fact that man can make his selection with much greater care. Man is able with the greatest nicety to pick out individuals from a large number, drop the others, and to employ only the privileged beings for propagation, which is not the case in natural selection. In natural conditions, besides the privileged individuals which first succeed in propagating themselves, some few or many of the less distinguished individuals will propagate themselves by the side of the former. Moreover, man can prevent the crossing of the original and the new form, which in natural selection is often unavoidable. If such a crossing, that is, a sexual connection, of the new variety with the original forms takes place, the offspring thereby produced generally returns to the original character. In natural selection, such a crossing can be avoided only when the new variety by migration separates from the original and isolates itself.

Natural selection therefore acts much more slowly; it requires much longer periods than the artificial process of selection. But it is an essential consequence of this difference, that the product of artificial selection disappears much more easily, and that the new form returns rapidly to the earlier one, which is not the case in natural selection. The new species arising from natural selection maintain themselves much more permanently, and return much less easily to the original form, than is the case with products of artificial selection, and accordingly maintain themselves during a much longer time than the artificial races produced by man. But these are only subordinate differences, which are explained by the different conditions of

natural and artificial selection, and in reality are connected only with differences in the duration of time. The nature of the transformation and the
means by which it is produced are entirely the same in both artificial and
natural selection. (Gen. Morph. ii. 248.)

The thoughtless and narrow-minded opponents of Darwin are never tired
of asserting that his theory of selection is a groundless conjecture, or at least
an hypothesis which has yet to be proved. That this assertion is completely
unfounded, may be perceived even from the outlines of the doctrine of selection which have just been discussed. Darwin assumes no kind of unknown
forces of nature, nor hypothetical conditions, as the acting causes for the
transformation of organic forms, but solely and simply the universally recognized vital activities of all organisms, which we term *Inheritance* and *Adaptation*. Every naturalist acquainted with physiology knows that these two
phenomena are directly connected with the functions of propagation and nutrition, and, like all other phenomena of life, are purely mechanical processes
of nature, that is, they depend upon the molecular phenomena of motion in
organic matter. That the interaction of these two functions effect a continual,
slow transmutation of organic forms, is a necessary result of the struggle
for existence. But this, again, is no more a hypothetical relation, nor one
requiring a proof, than is the interaction of Inheritance and Adaptation. The
struggle for life is a mathematical necessity, arising from the disproportion
between the limited number of places in nature's household, and the excessive number of organic germs. The origin of new species is moreover greatly
favoured by the active or passive *migrations* of animals and plants, which
takes place everywhere and at all times, without being, however, entitled to
rank as *necessary* agents in the process of natural selection.

The origin of new species by natural selection, or, what is the same thing,
by the interaction of Inheritance and Adaptation in the struggle for life, is
therefore a *mathematical necessity* of nature which needs no further proof.
Whoever, in spite of the present state of our knowledge, still seeks for *proofs*
for the Theory of Selection, only shows that he either does not thoroughly
understand the theory, or is not sufficiently acquainted with the biological
facts—has not the requisite amount of experimental knowledge in Anthropology, Zoology, and Botany.

If, as we maintain, natural selection is the great active cause which has
produced the whole wonderful variety of organic life on the earth, all the interesting phenomena of *human life* must also be explicable from the same
cause. For man is after all only a most highly-developed vertebrate animal,
and all aspects of human life have their parallels, or, more correctly, their
lower stages of development in the animal kingdom. The whole history of nations, or what is called "Universal History," must therefore be explicable by
means of "natural selection,"—must be a physico-chemical process, depending
upon the interaction of Adaptation and Inheritance in the struggle for life.
And this is actually the case. We shall give further proofs of this later on.

It appears of interest here to remark that not only *natural* selection, but also *artificial* selection exercises its influence in many ways in universal history. A remarkable instance of *artificial selection in man*, on a great scale, is furnished by the ancient Spartans, among whom, in obedience to a special law, all newly-born children were subject to a careful examination and selection. All those that were weak, sickly, or affected with any bodily infirmity, were killed. Only the perfectly healthy and strong children were allowed to live, and they alone afterwards propagated the race. By this means, the Spartan race was not only continually preserved in excellent strength and vigour, but the perfection of their bodies increased with every generation. No doubt the Spartans owed their rare degree of masculine strength and rough heroic valour (for which they are eminent in ancient history) in a great measure to this artificial selection.

Many tribes also among the Red Indians of North America (who at present are succumbing in the struggle for life to the superior numbers of the white intruders, in spite of a most heroic and courageous resistance) owe their rare degree of bodily strength and warlike bravery to a similar careful selection of the newly-born children. Among them, also, all children that are weak or affected with any infirmity are immediately killed, and only the perfectly strong individuals remain in life, and propagate the race. That the race becomes greatly strengthened, in the course of very many generations, by this artificial selection cannot in itself be doubted, and is sufficiently proved by many well known facts.

The opposite of this artificial selection of the wild Redskins and the ancient Spartans is seen in the individual selection which is universally practised in our modern military states, for the purpose of maintaining standing armies, and which, under the name of *military selection*, we may conveniently consider as a special form of selection. Unfortunately, in our day, militarism is more than ever prominent in our so-called "civilization"; all the strength and all the wealth of flourishing civilized states are squandered on its development; whereas the education of the young, and public instruction, which are the foundations of the true welfare of nations and the ennobling of humanity, are neglected and mismanaged in a most pitiable manner. And this is done in states which believe themselves to be the privileged leaders of the highest human intelligence, and to stand at the head of civilization. As is well known, in order to increase the standing army as much as possible, all healthy and strong young men are annually selected by a strict system of recruiting. The stronger, healthier, and more spirited a youth is, the greater is his prospect of being killed by needle-guns, cannons, and other similar instruments of civilization. All youths that are unhealthy, weak, or affected with infirmities, on the other hand, are spared by the "military selection," and remain at home during the war, marry, and propagate themselves. The more useless, the weaker, or infirmer the youth is, the greater is his prospect of escaping the recruiting officer, and of founding a family. While the healthy flower of youth dies on the battle-field, the feeble remainder enjoy the satisfaction of reproduction and of transmitting all their weaknesses and infirmities to their descendants. According to the laws of transmission by inheritance, there must necessarily follow in each succeeding generation, not only a further extension, but also a more deeply-seated development of weakness of body, and what is inseparable from it, a condition of mental weakness also. This and other forms of artificial selection practised in our civilized states sufficiently explain the sad fact that, in reality, weakness of the body and weakness of character are on the perpetual increase among civilized nations, and that, together with strong, healthy bodies, free and independent spirits are becoming more and more scarce.

To the increasing enervation of modern civilized nations, which is the necessary consequence of military selection, there is further added another evil. The progress of modern medical science, although still little able really to cure diseases, yet possesses and practises more than it used to do the art of prolonging life during lingering, chronic diseases for many years. Such ravaging evils as consumption, scrofula, syphilis, and also many forms of mental disorders, are transmitted by inheritance to a great extent, and transferred by sickly parents to some of their children, or even to the whole of their descendants. Now, the longer the diseased parents, with medical assistance, can drag on their sickly existence, the more numerous are the descendants who will inherit incurable evils, and the greater will be the number of individuals, again, in the succeeding generation, thanks to that artificial "*medical selection,*" who will be infected by their parents with lingering, hereditary disease.

If any one were to venture the proposal, after the examples of the Spartans and Redskins, to kill, immediately upon their birth, all miserable, crippled children to whom with certainty a sickly life could be prophesied, instead of keeping them in life injurious to them and to the race, our so-called "humane civilization" would utter a cry of indignation. But the same "humane civilization" thinks it quite as it should be, and accepts without a murmur, that at the outbreak of every war (and in the present state of civilized life, and in the continual development of standing armies, wars must naturally become more frequent) hundreds and thousands of the finest men, full of youthful vigour, are sacrificed in the hazardous game of battles. The same "humane civilization" at present praises the abolition of capital punishment as a "liberal measure!" And yet capital punishment for incorrigible and degraded criminals is not only just, but also a benefit to the better portion of mankind; the same benefit is done by destroying luxuriant weeds, for the prosperity of a well cultivated garden. As by a careful rooting out of weeds, light, air, and ground is gained for good and useful plants, in like manner, by the indiscriminate destruction of all incorrigible criminals, not only would the struggle for life among the better portion of mankind be made easier, but also an advantageous artificial process of selection would be set in practice, since the possibility of transmitting their injurious qualities by inheritance would be taken from those degenerate outcasts.

Against the injurious influence of artificial military and medical selection, we fortunately have a salutary counterpoise, in the invincible and much more powerful influence of *natural selection*, which prevails everywhere. For in the life of man, as well as in that of animals and plants, this influence is the most important transforming principle, and the strongest lever for progress and amelioration. The result of the struggle for life is that, in the long run, that which is better, because more perfect, conquers that which is weaker and imperfect. In human life, however, this struggle for life will ever become more and more of an intellectual struggle, not a struggle with weapons of murder. The organ which, above all others, in man becomes more perfect by the ennobling influence of natural selection, is the *brain*. The man with the most perfect understanding, not the man with the best revolver, will in the long run be victorious; he will transmit to his descendants the qualities of the brain which assisted him in the victory. Thus then we may justly hope, in spite of all the efforts of retrograde forces, that the progress of mankind towards freedom, and thus to the utmost perfection, will, by the happy influence of natural selection, become more and more certain.

CHAPTER VIII.

TRANSMISSION BY INHERITANCE AND PROPAGATION.

Universality of Inheritance and Transmission by Inheritance.—Special Evidences of the same.—Human Beings with four, six, or seven Fingers and Toes.—Porcupine Men.—Transmission of Diseases, especially Diseases of the Mind.—Original Sin.—Hereditary Monarchies.—Hereditary Aristocracy.—Hereditary Talents and Mental Qualities.—Material Causes of Transmission by Inheritance.—Connection between Transmission by Inheritance and Propagation.—Spontaneous Generation and Propagation.—Nonsexual or Monogonous Propagation.—Propagation by Self-Division.—Monera and Amoebae.—Propagation by the formation of Buds, by the formation of Germ-Buds, by the formation of Germ-Cells.—Sexual or Amphigonous Propagation.—Formation of Hermaphrodites.—Distinction of Sexes, or Gonochorism.—Virginal Breeding, or Parthenogenesis.—Material Transmission of Peculiarities of both Parents to the Child by Sexual Propagation.—Difference between Transmission by Inheritance in Sexual and in Asexual Propagation.

THE reader has, in the last chapter, become acquainted with natrual selection according to Darwin's theory, as the constructive force of nature which produces the different forms of animal and vegetable species. By natural selection we understand the interaction which takes place in the struggle for life between the *transmission by inheritance* and the *mutability* of organisms, between two physiological functions which are innate in all animals and plants, and which may be traced to other processes of life—the functions of propagation and nutrition. All the different forms of organisms, which people are usually inclined to look upon as the products of a creative power, acting for a definite purpose, we, according to the Theory of Selection, can conceive as the necessary productions of natural selection, working without a purpose,—as the unconscious interaction between the two properties of Mutability and Hereditivity. Considering the importance which accordingly belongs to these vital properties of organisms, we must examine them a little more closely, and employ a chapter with the consideration of Transmission by Inheritance. (Gen. Morph. ii. 170–191.)

Strictly speaking, we must distinguish between Hereditivity (Transmissivity) and Inheritance (Transmission). Hereditivity is the power of transmission, the capability of organisms to transfer their peculiarities to their descendants by propagation. Transmission by Inheritance, or Inheritance simply, on the other hand, denotes the exercise of the capability, the actual transmission.

Hereditivity and Transmission by Inheritance are such universal, everyday phenomena, that most people do not heed them, and but few are inclined to reflect upon the operation and import of these phenomena of life. It is generally thought quite natural and self-evident that every organism should produce its like, and that children should more or less resemble their parents. Heredity is usually only taken notice of and discussed in cases relating to some special peculiarity, which appears for the first time in a human individual without having been inherited, and then is transmitted to his descendants. It shows itself in a specially striking manner in the case of certain diseases, and in unusual and irregular (monstrous) deviations from the usual formation of the body.

Among these cases of the inheritance of monstrous deviations, those are specially interesting which consist in an abnormal increase or decrease of the number five in the fingers or toes of man. It is not unfrequently observed in families through several generations, that individuals have six fingers on each hand, or six toes on each foot. Less frequent is the number of four or seven fingers or toes. The unusual formation arises at first from a single individual who, from unknown causes, is born with an excess of the usual number of fingers and toes, and transmits these, by inheritance, to a portion of his descendants. In one and the same family it has happened that, throughout three, four, or more generations, individuals have possessed six fingers and toes. In a Spanish family there were no less than forty individuals distinguished by this excess. The transmission of the sixth finger or toe is not permanent or enduring in all cases, because six-fingered people always intermarry again with those possessing five fingers. If a six-fingered family were to propagate by pure in-breeding, if six-fingered men were always to marry six-fingered women, this characteristic would become permanent, and a special six-fingered human race would arise. But as six-fingered men usually marry five-fingered women, and *vice versa*, their descendants for the most part show a very mixed numerical relation, and finally, after the course of some generations, revert again to the normal number of five. Thus, for example, among eight children of a six-fingered father and a five-fingered mother, two children may have on both hands and feet six fingers and toes, four children may have a mixed number, and two children may have the usual number of five on both hands and feet. In a Spanish family, each child except the youngest had the number six on both hands and feet; the youngest, only, had the usual number on both hands and feet, and the six-fingered father of the child refused to recognize the last one as his own.

The power of inheritance, moreover, shows itself very strikingly in the formation and colour of the human skin and hair. It is well known how

exactly the nature of the complexion in many families—for instance, a peculiar soft or rough skin, a peculiar luxuriance of the hair, a peculiar colour
and largeness of the eyes—is transmitted through many generations. In like
manner, peculiar local growths or spots on the skin, the so-called moles,
freckles, and other accumulations of pigment which appear in certain places, are frequently transmitted through several generations so exactly, that
in the descendants they appear on the same spots on which they existed in
the parents. The porcupine men of the Lambert family, who lived in London
last century, are especially celebrated. Edward Lambert, born in 1717, was
remarkable for a most unusual and monstrous formation of the skin. His
whole body was covered with a horny substance, about an inch thick, which
rose in the form of numerous thorn-shaped and scale-like processes, more
than an inch long. This monstrous formation of the outer skin, or epidermis, was transmitted by Lambert to his sons and grandsons, but not to his
granddaughters. The transmission in this instance remained in the male
line, as is often the case. In like manner, an excessive development of fat
in certain parts of the body is often transmitted only in the female line. I
scarcely need call to mind how exactly the characteristic formation of the
face is transmitted by inheritance; sometimes it remains within the male,
sometimes within the female line; sometimes it is blended in both.

The phenomena of transmission by inheritance of pathological conditions, especially of the different forms of human diseases, are very instructive and generally known. Diseases of the respiratory organs, the glands,
and of the nervous system, are specially liable to be transmitted by inheritance. Very frequently there suddenly appears in an otherwise healthy family a disease until then unknown among them; it is produced by external
causes, by conditions of life causing disease. This disease, brought about
in an individual by external cause, is propagated and transmitted to his
descendants, and some or all of them then suffer from the same disease. In
case of diseases of the lungs, for instance in consumption, this sad transmission by inheritance is well known, and it is the same with diseases of the
liver, with syphilis, and diseases of the mind. The latter are specially interesting. Just as peculiar characteristic features of man—pride, ambition,
frivolity, etc.—are transmitted to the descendants strictly by inheritance, so
too are the peculiar abnormal manifestations of mental activity, which are
usually called fixed ideas, despondency, imbecility, and generally "diseases
of the mind." This distinctly and irrefragably shows that the soul of man,
just as the soul of animals, is a purely mechanical activity, the sum of the
molecular phenomena of motion in the particles of the brain, and that it
is transmitted by inheritance, together with its substratum, just as every
other quality of the body is materially transmitted by propagation.

When this exceedingly important and undeniable fact is mentioned, it
generally causes great offence, and yet in reality it is silently and universally acknowledged. For upon what else do the ideas of "hereditary sin,"

"hereditary wisdom," and "hereditary aristocracy," etc., rest than upon the conviction that the *quality of the human mind* is transmitted by propagation—that is, by a purely *material* process—through the body, from the parents to the descendants? The recognition of this great importance of transmission by inheritance is shown in a number of human institutions, as for example, among many nations in the division into castes, such as the castes of warriors, castes of priests, and castes of labourers, etc. It is evident that the institution of such castes originally arose from the notion of the great importance of hereditary distinctions possessed by certain families, which it was presumed would always be transmitted by the parents to the children. The institution of an hereditary aristocracy and an hereditary monarchy is to be traced to the notion of such a transmission of special excellencies. However, it is unfortunately not only virtues, but also vices that are transmitted and accumulated by inheritance; and if, in the history of the world, we compare the different individuals of the different dynasties, we shall everywhere find a great number of proofs of the transmission of qualities by inheritance, but fewer of transmissions of virtues than of vices. Look only, for example, at the Roman emperors, at the Julii and the Claudii, or at the Bourbons in France, Spain, and Italy!

In fact, scarcely anywhere could we find such a number of striking examples of the remarkable transmission of bodily and mental features by inheritance, as in the history of the reigning houses in hereditary monarchies. This is specially true in regard to the diseases of the mind previously mentioned. It is in reigning families that mental disorders are hereditary in an unusual degree. Thus Esquirol, distinguished for his knowledge of mental diseases, proved that the number of insane individuals in the reigning houses was, in proportion to the number among the ordinary population, as 60 to 1; that is, that disorders of the brain occur 60 times more frequently in the privileged families of the ruling houses than among ordinary people. If equally accurate statistics were made of the hereditary nobility, the result would probably be that here also we should find an incomparably larger contingent of mental diseases than among the common, ignoble portion of mankind. This phenomenon can scarcely astonish us if we consider what injury these privileged castes inflict upon themselves by their unnatural, one-sided education, and by their artificial separation from the rest of mankind. By this means many dark sides of human nature are specially developed and, as it were, artificially bred, and, according to the laws of transmission by inheritance, are propagated through series of generations with ever-increasing force and dominance.

It is sufficiently obvious from the history of nations how in successive generations of many dynasties, for example, of the princes of Saxon Thuringia and of the Medici, the noble solicitude for the most perfect human accomplishments in science and art were retained and transmitted from father to son; and how, on the other hand, in many other dynasties, for centuries a special partiality for the profession of war, for the oppression of human freedom, and for other rude acts of violence, have been hereditary. In like manner talents for special mental activities are transmitted in many families for generations, as, for instance, talent for mathematics, poetry, music, sculpture, the investigation of nature, philosophy, etc. In the family of Bach there have been no less than twenty-two eminent musicians. Of course the transmission of such peculiarities of mind depends upon the material process of reproduction, as does the transmission of mental qualities in general. In this case again, the vital phenomenon, the manifestation of force (as everywhere in nature), is directly connected with definite relations in the admixture of the material components of the organism. It is this definite proportion and molecular motion of matter which is transmitted by generation.

Now, before we examine the numerous, and in some cases most interesting and important, laws of transmission by inheritance, let us make ourselves acquainted with the actual nature of the process. The phenomena of transmission by inheritance are generally looked upon as something quite mysterious, as peculiar processes which cannot be fathomed by natural science, and the causes and actual nature of which cannot be understood. It is precisely in such a case that people very generally assume supernatural influences. But even in the present state of our physiology it can be proved with complete certainty that all the phenomena of inheritance are entirely natural processes, that they are produced by mechanical causes, and that they depend on the material phenomena of motion in the bodies of organisms, which we may consider as a part of the phenomena of propagation. All the phenomena of Heredity and the laws of Transmission by Inheritance can be traced to the material process of *Propagation*.

Every organism, every living individual, owes its existence *either* to an act of unparental or *Spontaneous Generation* (Generatio Spontanea, Archigonia), or to an act of Parental Generation or *Propagation* (Generatio Parentalis, Tocogonia). In a future chapter we shall have to consider Spontaneous Generation, or Archigony. At present we must occupy ourselves with Propagation, or Tocogony, a closer examination of which is of the utmost importance for understanding transmission by inheritance. Most of my readers probably only know those phenomena of Propagation which are seen universally in the higher plants and animals, the processes of Sexual Propagation, or Amphigony. The processes of Non-sexual Propagation, or Monogony, are much less generally known. The latter, however, are far more suited to throw light upon the nature of transmission by inheritance in connection with propagation.

For this reason, we shall first consider only the phenomena of *non-sexual* or *monogonic propagation* (Monogonia). This appears in a variety of different forms, as for example, self-division, formation of buds, the formation of germ-cells or spores (Gen. Morph. ii. 36–58). It will be most instructive, first, to examine the propagation of the simplest organisms known to us, which we shall have to return to later, when considering the question of spontaneous generation. These very simplest of all organisms yet known, and which, at the same time, are the simplest imaginable organisms, are the *Monera* living in water; they are very small living corpuscles, which, strictly speaking, do not at all deserve the name of organism. For the designation "organism," applied to living creatures, rests upon the idea that every living natural body is composed of organs, of various parts, which fit into one another and work together (as do the different parts of an artificial machine), in order to produce the action of the whole. During late years we have become acquainted with *Monera*, organisms which are, in fact, not composed of any organs at all, but consist entirely of shapeless, simple, homogeneous matter. The entire body of one of these Monera, during life, is nothing more than a shapeless, mobile, little lump of mucus or slime, consisting of an albuminous combination of carbon. Simpler or more imperfect organisms we cannot possibly conceive.

The first complete observations on the natural history of a Moneron (*Protogenes primordialis*) were made by me at Nice, in 1864. Other very remarkable Monera I examined later (1866) in Lanzarote, one of the Canary Islands, and in 1867 in the Straits of Gibraltar. The complete history of one of these Monera, the orange-red *Protomyxa aurantiaca*, is represented in Plate I, and its explanation is given in the Appendix. I have found some curious Monera also in the North Sea, off the Norwegian coast, near Bergen. Cienkowski has described (1865) an interesting Moneron from fresh waters, under the name of *Vampyrella*. But perhaps the most remarkable of all Monera was discovered by Huxley, the celebrated English zoologist, and called *Bathybius haeckelii*. "Bathybius" means, living in the deep. This wonderful organism lives in immense depths of the ocean, which are over 12,000—indeed, in some parts 24,000 feet below the surface, and which have become known to us within the last ten years, through the laborious investigations made by the English. There, among the numerous Polythalamia and Radiolaria which inhabit the fine calcareous mud of these abysses, the *Bathybius* is found in great quantities, sometimes in the shape of roundish, formless lumps of mucus, sometimes in the form of a network of mucus, covering fragments of stone and other objects. Small particles of chalk are frequently embedded in these mucous gelatinous masses, and are, perhaps, products of their secretion. The entire body of this remarkable *Bathybius* consists solely of shapeless plasma, or protoplasm, as in the case of the other Monera—that is, it consists of the same albuminous combination of carbon, which in infinite modifications is found in all organisms, as the essential and never-failing seat of the phenomena of life. I have given a detailed description and drawing of the *Bathybius* and other Monera in my "Monographie der Moneren," 1870,[15] from which the drawing in Fig. 9 is taken.

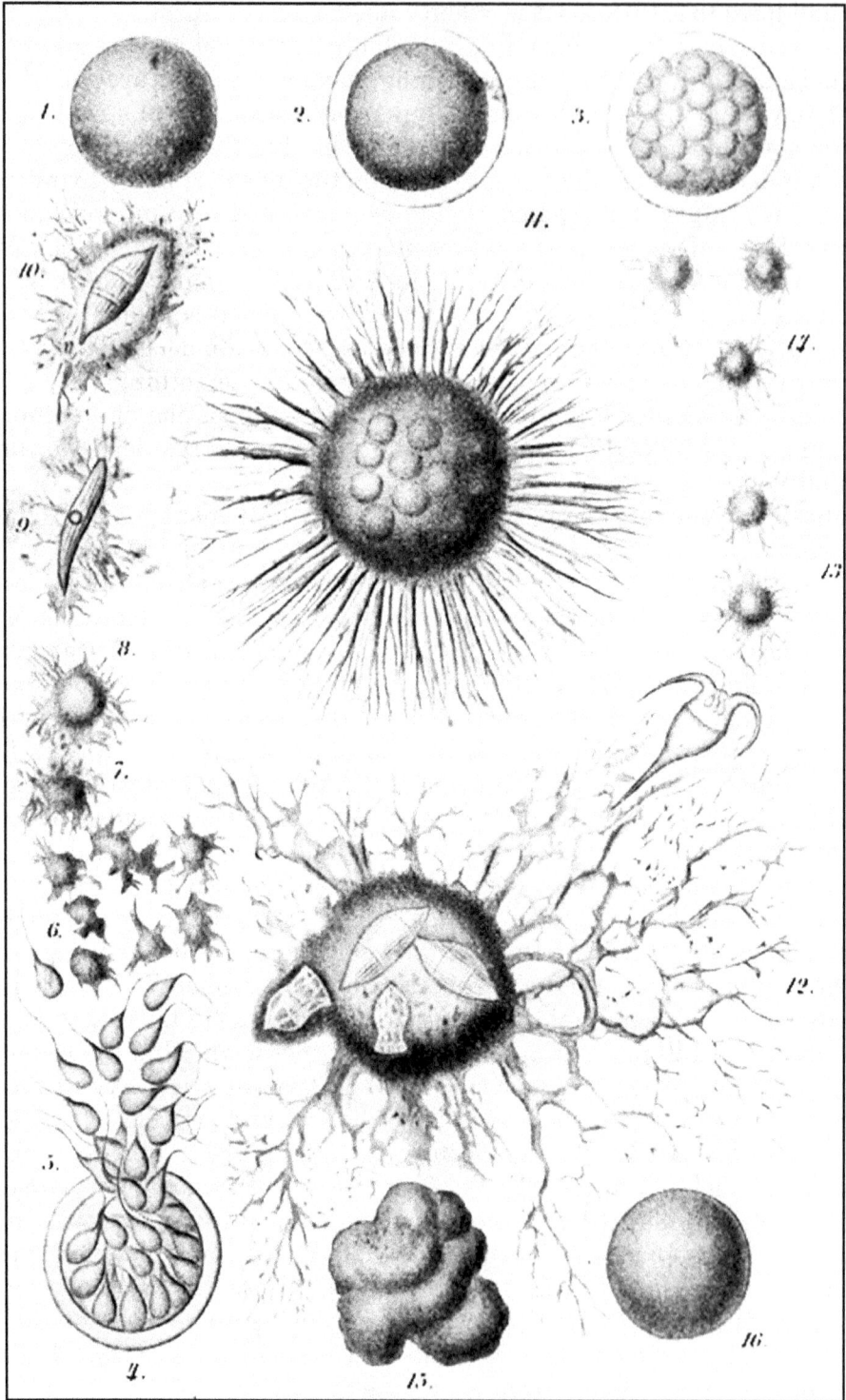

Protomyxa aurantiaca.

In a state of rest most Monera appear as small globules of mucus or slime, invisible, or nearly so, to the naked eye; they are at most as large as a pin's head. When the Moneron moves itself, there are formed on the upper surface of the little mucous globule, shapeless, fingerlike processes, or very fine radiated threads; these are the so-called false feet, or pseudopodia. The false feet are simple, direct continuations of the shapeless albuminous mass, of which the whole body consists. We are unable to perceive different parts in it, and we can give a direct proof of the absolute simplicity of the semi-fluid mass of albumen, for with the aid of the microscope we can follow the Moneron as it takes in nourishment. When small particles suited for its nourishment—for instance, small particles of decayed organic bodies or microscopic plants and infusoria—accidentally come into contact with the Moneron, they remain hanging to the sticky semi-fluid globule of mucus, and here create an irritation, which is followed by a strong afflux of the mucous substance, and, in consequence, they become finally completely inclosed by it, or are drawn into the body of the Moneron by displacement of the several albuminous particles, and are there digested, being absorbed by simple diffusion (endosmosis).

Just as simple as the process of nutrition is the *propagation* of these primitive creatures, which in reality we can neither call animals nor plants. All Monera propagate themselves only in an asexual manner by monogony; and in the simplest case, by that kind of monogony which we place at the head of the different forms of propagation, that is, by self-division. When such a little globule, for example a Protamoeba or a Protogenes, has attained a certain size by the assimilation of foreign albuminous matter, it falls into two pieces; a pinching in takes place, contracting the middle of the globule on all sides, and finally leads to the separation of the two halves (compare Fig. 1). Each half then becomes rounded off, and now appears as an independent individual, which commences anew the simple course of the vital phenomena of nutrition and propagation. In other Monera (*Vampyrella*),

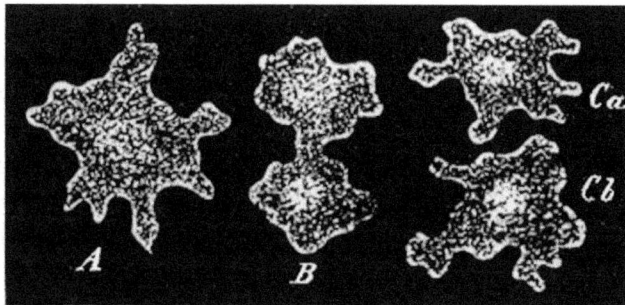

FIG. 1.—Propagation of the simplest organism, a Moneron, by self-division. *A.* The entire Moneron, a Protamoeba. *B.* It falls into two halves by a contraction in the middle. *C.* Each of the two halves has separated from the other, and now represents an independent individual.

the body in the process of propagation does not fall into two, but into four equal pieces, and in others, again (*Protomonas, Protomyxa, Myxastrum*), at once into a number of small globules of mucus, each of which again, by simple growth, becomes like the parent body. Here it is evident that the process of *propagation is nothing but a growth of the organism beyond its own individual limit of size.*

The simple method of propagation of the Moneron by self-division is, in reality, the most universal and most widely spread of all the different modes of propagation; for by the same simple process of division, *cells* also propagate themselves. Cells are those simple organic individuals, a large number of which constitute the bodies of most organisms, the human body not excepted. With the exception of the organisms of the lowest order, which have not even the perfect form of a cell (Monera), or during life only represent a single cell (many Protista and single-celled plants), the body of every organic individual is composed of a great number of cells. Every organic cell

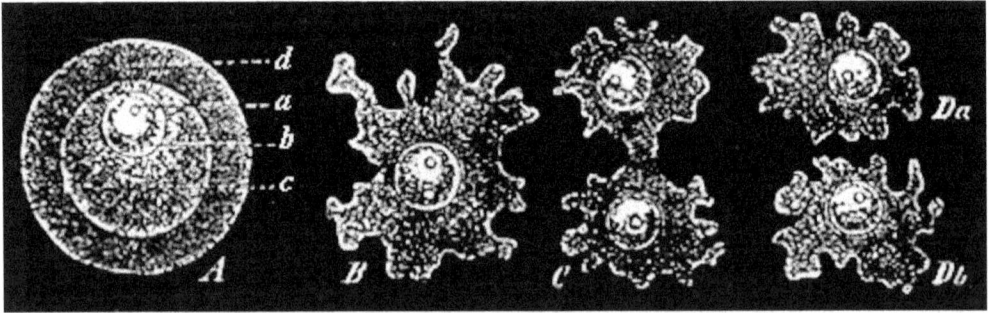

Fig. 2.—Propagation of a single-celled organism, *Amoeba sphaerococcus*, by self-division. *A.* The enclosed Amoeba, a simple globular cell consisting of a lump of protoplasm (*c*), which contains a kernel (*b*) and a kernel speck (*a*), and is surrounded by a cell-membrane or capsule. *B.* The free Amoeba, which has burst and left the cyst or cell-membrane. *C.* It begins to divide by its kernel forming two kernels, and by the cell-substance between the two becoming contracted. *D.* The division is completed by the cell-substance likewise falling into two halves (*Da* and *Db*).

is to a certain degree an independent organism, a so-called "elementary organism," or an "individual of the first order." Every higher organism is, in a measure, a society or a state of such variously shaped elementary individuals, variously developed by division of labour.[39] Originally every organic cell is only a single globule of mucus, like a Moneron, but differing from it in the fact that the homogeneous albuminous substance has separated itself into two different parts, a firmer albuminous body, the *cell-kernel* (nucleus), and an external, softer albuminous body, the *cell-substance* or *body* (protoplasma). Besides this, many cells later on form a third (frequently absent) distinct part, inasmuch as they cover themselves with a capsule, by exuding an outer pellicle or *cell-membrane* (membrana). All other forms of cells, besides these, are of subordinate importance, and are of no further interest to us here.

Every organism composed of many cells was originally a single cell, and it becomes many-celled owing to the fact that the original cell propagates itself by self-division, and that the new individual cells originating in this manner remain together, and by division of labour form a community or a state. The forms and vital phenomena of all many-celled organisms are merely the effect or the expression of all the forms and vital phenomena of all the individual cells of which they are composed. The egg, from which most animals and plants are developed, is a simple cell.

The single-celled organisms, that is, those which during life retain the form of a single cell, for example the Amoebae, as a rule propagate themselves in the simplest way by self-division. This process differs from the previously described self-division of the Moneron only in the fact that at the commencement the firmer cell-kernel (nucleus) falls into two halves, by a pinching in at its middle. The two young kernels separate from each other and act now as two distinct centres of attraction upon the surrounding softer albuminous matter, that is, the cell-substance (protoplasma). By this process finally the latter also divides into two halves, and there now exist two new cells, which are like the mother cell. If the cell was surrounded by a membrane, this either does not divide at all, as in the case of egg-cleavage (Fig. 3, 4), or it passively follows the active pinching in of the protoplasm; or, lastly, every new cell exudes a new membrane for itself.

The non-independent cells which remain united in communities or states, and thus constitute the body of higher organisms, are propagated in the same manner as are independent single-celled organisms, for example, *Amoeba* (Fig. 2). Just as in that case, the cell with which most animals and plants commence their individual existence, namely, the egg, multiplies itself by simple division. When an animal, for instance a mammal (Fig. 3, 4), develops out of an egg, this process of development always begins by the simple egg-cell (Fig. 3) forming an accumulation of cells (Fig. 4) by continued self-division. The outer covering, or cell membrane, of the globular egg remains undivided. First, the cell-kernel of the egg (the so-called germinal vesicle) divides itself into two kernels, then follows the cell-substance (the yolk of the egg) (Fig. 4 A). In like manner, the two cells, by continued self-division, separate into four (Fig. 4 B), these into eight (Fig. 4 C), into sixteen,

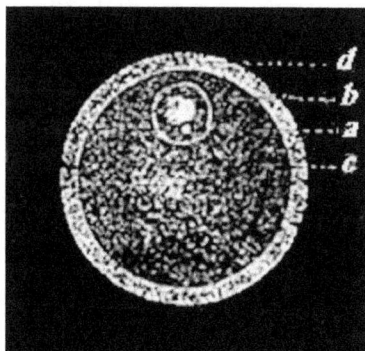

FIG. 3.—Egg of a mammal (a simple cell). *a.* The small kernel speck or nucleolus (the so-called germ-spot of the egg). *b.* Kernel or nucleus (the so-called germ-bladder of the egg). *c.* Cell-substance or protoplasm (the so-called yolk of the egg). *d.* Cell-capsule or membrane (membrane of the yolk) of the egg; called in mammals, on account of its transparency, Membrana pellucida.

thirty-two, etc., and finally there is produced a globular mass of very numerous little cells (Fig. 4 *D*). These now, by further increase and heterogeneous
development (division of labour), gradually build up the compound many-
celled organism. Every one of us, at the commencement of our individual
development, has undergone the very same process as that represented in
Fig. 4. The egg of a mammal—represented in Fig. 3, and its development in
Fig. 4—might as well be that of a man, as of an ape, dog, horse, or any other
placental mammal.

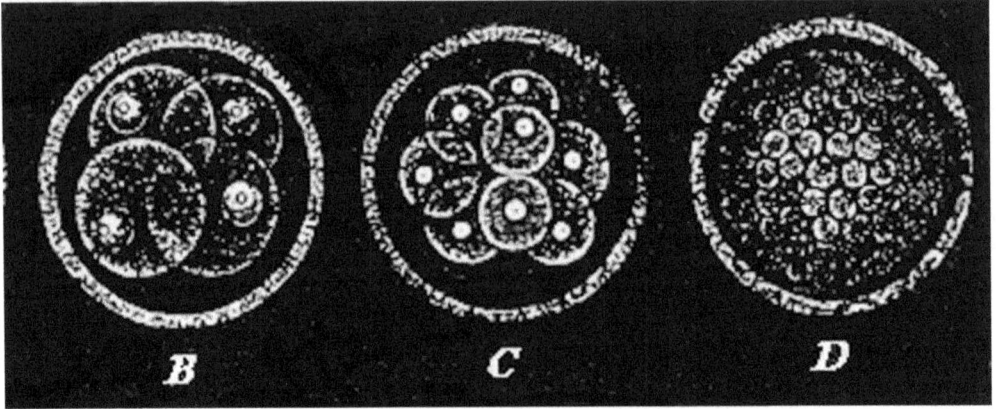

FIG. 4.—First commencement of the development of a mammal's egg, the so-called
"cleavage of the egg" (propagation of the egg-cell by repeated self-division). *A*. The egg, by the
formation of the first furrow, falls into two cells. *B*. These separate by division into four cells.
C. The latter have divided into eight cells. *D*. By repeated division a globular accumulation
of numerous cells has arisen.

Now, when one examines this simplest form of propagation, this self-
division, it surely cannot be considered wonderful that the products of the
division of the original organism should possess the same qualities as the
parental individual. For they are parts or halves of the parental organism,
and the matter or substance in both halves is the same, and as both the
young individuals have received an equal amount and the same quality of
matter from the parent individual, one can but consider it natural that the
vital phenomena, the physiological qualities should be the same in both
children. In fact, in regard to their form and substance, as well as to their
vital phenomena, the two produced cells can in no respect be distinguished
from one another, or from the mother cell. They have *inherited* from her the
same nature.

But this same simple propagation by self-division is not only confined to
simple cells—it is the same also in the higher many-celled organisms; for example, in the coral zoophytes. Many of them which exhibit a high complexity of composition and organization, nevertheless, propagate themselves by
simple division. In this case the whole organism, with all its organs, falls
into two equal halves as soon as by growth it has attained a certain size.

Each half again develops itself, by growth, into a complete individual. Here, again, it is surely self-evident that the two products of division will share the qualities of the parental organism, as they themselves are in fact halves of that parent.

Next to propagation by division we come to propagation by the *formation of buds*. This kind of monogony is exceedingly widely spread. It occurs both in the case of simple cells (though not frequently) and in the higher organisms composed of many cells. The formation of buds is universal in the vegetable kingdom, less frequent in the animal kingdom. However, here also it occurs in the tribe of Plant-like Animals, especially among the Coral Zoophytes, and among the greater portion of the Hydroid Polyps very frequently, further also among some worms (Planarian Worms, Ring-Worms, Moss Animals, Tunicates). Most branching animal-trees or colonies, which are exceedingly like branching plants, arise like those plants, by the formation of buds.

Propagation by the *formation of buds* (Gemmatio) is essentially distinguished from propagation by division, in the fact that the two organisms thus produced by budding are not of equal age, and therefore at first are not of equal value, as they are in the case of division. In division we cannot clearly distinguish either of the two newly produced individuals as the parental, that is as the producer, because, in fact, both have an equal share in the composition of the original parental individual. If, on the other hand, an organism sends out a bud, then the latter is the child of the former. The two individuals are of unequal size and of unequal form. If, for instance, a cell propagates itself by the formation of buds, we do not see the cell fall into two equal halves, but there appears at one point of it a protuberance, which becomes larger and larger, more or less separates itself from the parental cell, and then grows independently. In like manner we observe in the budding of a plant or animal, that a small local growth arises on a part of the mature individual, which growth becomes larger and larger, and likewise more or less separates itself from the parental organism by an independence in its growth. The bud, after it has attained a certain size, may either completely separate itself from the parental individual, or it may remain connected with it and form a stock or colony, whilst at the same time its life may be quite independent of that of its parent. While the growth which starts the propagation, in the case of self-division, is a total one affecting the whole body, it is in the formation of buds only partial, affecting merely a portion of the parental organism. But here, also, the bud—the newly-produced individual which remains so long most directly connected with the parental organism, and which proceeds from it—retains the essential qualities and the original tendency of development of its parent.

A third mode of non-sexual propagation, that of the *formation of germ-buds* (Polysporogonia), is intimately connected with the formation of buds. In the case of the lower, imperfect organisms, among animals, especially in the case of the Plant-like animals and Worms, we very frequently find that in the interior of an individual composed of many cells, a small group of cells separates itself from those surrounding it, and that this small isolated group gradually develops itself into an individual, which, becomes like the parent, and sooner or later comes out of it. Thus, for example, in the body of the Fluke-worms (Trematodes) there often arise numerous little bodies consisting of many cells, that is *germ-buds*, or *polyspores*, which, at an early stage separate themselves completely from the parent body, and leave it when they have attained a certain stage of development.

The formation of germ-buds is evidently but little different from real budding. But, on the other hand, it is connected with a fourth kind of non-sexual propagation, which almost forms a transition to sexual reproduction, namely, the *formation of germ-cells* (Monosporogonia), which is often briefly called formation of spores (sporogonia). In this case it is no longer a group of cells, but a single cell, which separates itself from the surrounding cells in the interior of the producing organism, and which only becomes further developed after it has come out of its parent. After this *germ-cell*, or monospore (or, briefly, spore), has left the parental individual, it multiplies by division, and thus forms a many-celled organism, which by growth and gradual development attains the hereditary qualities of the parental organism. This occurs very generally among lower plants (Cryptogama).

Although the formation of germ-cells very much resembles the formation of germ buds, it evidently and very essentially differs from the latter, and also from the other forms of non-sexual propagation which have previously been mentioned, by the fact that only a very small portion of the producing organism takes part in the propagation and, accordingly, in the transmission by inheritance. In the case of self-division, where the whole organism falls into two halves, in the formation of buds, where a considerable portion of the whole body, already more or less developed, separates from the producing individual, we easily understand that the forms and vital phenomena should be the same in the producing and produced organism. It is much more difficult to understand in the formation of germ-buds, and more difficult still in the formation of germ-cells, how this very small, quite undeveloped portion of the body, this group of cells, or this single cell, not only directly takes with it certain parental qualities into its independent existence, but also after its separation from the parental individual develops into a many-celled body, and in this repeats the forms and vital phenomena of the original producing organism. This last form of monogonic propagation—that of the germ cells, or spore-formation—leads us directly to a form of propagation which is the most difficult of all to explain, namely, sexual propagation.

Sexual or amphigonic propagation (Amphigonia) is the usual method of propagation among all higher animals and plants. It is evident that it has only developed, at a very late period of the earth's history, from non-sexual propagation, and apparently in the first instance from the method of propagation by germ-cells. In the earliest periods of the organic history of the earth, all organisms propagated themselves in a non-sexual manner, as numerous lower organisms still do, especially all those which are at the lowest stage of organization, and which, strictly speaking, can be considered neither as animals nor as plants, and which therefore, as primary creatures, or Protista, are best excluded from both the animal and vegetable kingdoms. In the case of the higher animals and plants, the increase of individuals, as a rule, is at present brought about in the majority of cases by sexual propagation.

In all the chief forms of non-sexual propagation mentioned above—in fission, in the formation of buds, germ buds, and germ cells—the separated cell or group of cells was able by itself to develop into a new individual, but in the case of sexual propagation the cell must first be fructified by another generative substance. The fructifying male sperm must first mix with the female germ-cell (the egg) before the latter can develop into a new individual. These two different generative substances, the male sperm and the female egg, are either produced by one and the same individual hermaphrodite (Hermaphroditismus), or by two different individuals (sexual separation, Gonochorismus) (Gen. Morph. ii. 58, 59).

The simpler and more ancient form of sexual propagation is through double-sexed individuals (Hermaphroditismus). It occurs in the great majority of plants, but only in a minority of animals, for example, in the garden snails, leeches, earth-worms, and many other worms. Every single individual among hermaphrodites produces within itself materials of both sexes—eggs and sperm. In most of the higher plants every blossom contains both the male organ (stamens and anther) and the female organs (style and germ). Every garden snail produces in one part of its sexual gland eggs, and in another part sperm. Many hermaphrodites can fructify themselves; in others, however, copulation and reciprocal fructification of both hermaphrodites is necessary for causing the development of the eggs. This latter case is evidently a transition to sexual separation.

Sexual separation (Gonochorismus,) which characterizes the more complicated of the two kinds of sexual reproduction, has evidently been developed from the condition of hermaphroditism at a late period of the organic history of the world. It is at present the universal method of propagation of the higher animals, and occurs, on the other hand, only in the minority of plants (for example, in many aquatic plants, *e.g. Hydrocharis, Vallisneria;* and in trees, *e.g.* Willows, Poplars). Every organic individual, as a non-hermaphrodite (Gonochoristus), produces within itself only one of two generative substances, either the male or the female. The female individuals, both in animals and plants, produce eggs or egg-cells. The eggs of plants in the case of flowering plants (Phanerogama), are commonly called "embryo sacs"; in the case of flowerless plants (Cryptogama), "fruit spores." In animals, the male individual secretes the fructifying sperm (sperma); in plants, the corpuscles, which correspond to the sperm. In the Phanerogama, these are the pollen grains, or flower-dust; in the Cryptogama, a sperm, which, like that of most animals, consists of floating vibratile cells actively moving in a fluid—the zoosperms, spermatozoa, or sperm-cells.

The so-called *virginal reproduction* (Parthenogenesis) offers an interesting form of transition from sexual reproduction to the non-sexual formation of germ-cells (which most resembles it); it has been demonstrated to occur in many cases among Insects, especially by Siebold's excellent investigations. In this case germ-cells, which otherwise appear and are formed exactly like egg-cells, become capable of developing themselves into new individuals without requiring the fructifying seed. The most remarkable and most instructive of the different partheno-genetic phenomena are furnished by those cases in which the same germ-cells, according as they are fructified or not, produce different kinds of individuals. Among our common honey bees, a male individual (a drone) arises out of the eggs of the queen, if the egg has not been fructified; a female (a queen, or working bee), if the egg has been fructified. It is evident from this, that in reality there exists no wide chasm between sexual and non-sexual reproduction, but that both modes of reproduction are directly connected. The parthenogenesis of Insects must probably be regarded as a *relapse* from the sexual mode of propagation (possessed by the original parents of the insects) to the earlier condition of non-sexual propagation. (Gen. Morph. ii. 86.) In any case, however, sexual reproduction, both in plants and animals, which seems such a wonderful process, has only arisen at a later date out of the more ancient process of non-sexual reproduction. In both cases heredity is a necessary part of the phenomenon.

In all the different modes of propagation the essential point of the process is invariably a detachment of a portion of the parental organism possessing the capability of leading an individual, independent existence. We may, therefore, in all cases expect, *à priori*, that the produced individuals—which are, in fact, as is commonly said, "the flesh and blood" of the parents—will receive the vital characteristics and qualities of form which the parental individuals possess. It is simply a larger or smaller quantity of the parental material, in fact of its albuminous protoplasm, or cell-substance, which passes to the produced individual. But together with the material, its vital properties—that is, the molecular motions of the plasma—are transmitted, which then manifest themselves in its form. Inheritance by sexual breeding loses very much of the mysterious and wonderful character which it at first sight possesses for the uninitiated, if we consider the above-mentioned series of the different modes of propagation, and their connection one with another. It at first appears exceedingly wonderful that in the sexual propagation of man, and of all higher animals, the small egg, the minute cell, often invisible to the naked eye, is able to transfer to the produced organism all the qualities of the maternal organism, and, no less mysterious, that at the same time the essential qualities of the paternal organism are transferred to the offspring by means of the male sperm, which fructifies the egg-cell by means of a viscid substance in which minute thread-like cells or zoosperms move about. But as soon as we compare the connected stages of the different kinds of propagation, in which the produced organism separates itself more and more as a distinct growth from the parental individual, and more or less early enters upon its independent career; as soon as we consider, at the same time, that the growth and development of every higher organism only depends upon the increase of the cells composing it—that is, upon their simple propagation by division—it becomes quite evident that all these remarkable processes belong to one series.

The life of every organic individual is nothing but a connected chain of very complicated material phenomena of motion. These motions must be considered as changes in the position and combination of the molecules, that is, of the smallest particles of animated matter (of atoms placed together in the most varied manner). The specific, definite tendency of these orderly, continuous, and inherent motions of life depends, in every organism, upon the chemical mingling of the albuminous generative matter to which it owes its origin. In man, as in the case of the higher animals which propagate themselves in a sexual manner, the individual vital motion commences at the moment in which the egg-cell is fructified by the spermatic filaments of the seed, in which process both generative substances actually mix; and here the tendency of the vital motion is determined by the specific, or more accurately, by the individual nature of the sperm as well as of the egg. There can be no doubt as to the purely mechanical material nature of this process. But here we stand full of wonder and astonishment before the infinite and inconceivable delicacy of this albuminous matter. We are amazed at the undeniable fact that the simple egg-cell of the maternal organism, and a single paternal sperm-thread, transfer the molecular individual vital motion of these two individuals to the child so accurately, that afterwards the minutest bodily and mental peculiarities of both parents reappear in it.

Here we stand before a mechanical phenomenon of nature of which Virchow, whose genius founded the "cellular pathology," says with full justice: "If the naturalist cared to follow the custom of historians and preachers, and to clothe phenomena, which are in their way unique, with the hollow pomp of ponderous and sounding words, this would be the opportunity for him; for we have now approached one of those great mysteries of animal nature, which encircle the region of animal life as opposed to all the rest of the world of phenomena. The question of the formation of cells, the question of the excitation of a continuous and equable motion, and, finally, the questions of the independence of the nervous system and of the soul—these are the great problems on which the human mind can measure its strength." To comprehend the relation of the male and female to the egg-cell is almost as much as to solve all those mysteries. The origin and development of the egg-cell in the mother's body, the transmission of the bodily and mental peculiarities of the father to it by his seed, touch upon all the questions which the human mind has ever raised about man's existence. And, we add, these most important questions are solved, by means of the Theory of Descent, in a purely mechanical and purely monistic sense!

There can then be no further doubt that, in the sexual propagation of man and all higher organisms, inheritance, which is a purely mechanical process, is directly dependent upon the material continuity of the producing and produced organism, just as is the case in the simplest non-sexual propagation of the lower organisms. However, I must at once take this opportunity of drawing attention to an important difference which inheritance presents in sexual and non-sexual propagation. It is a fact long since acknowledged, that the individual peculiarities of the producing organism are much more accurately transmitted to the produced organism by non-sexual than by sexual propagation. Gardeners have for a long time made use of this fact in many ways. When, for instance, a single individual of a species of tree with stiff, upright branches accidentally produces down-hanging branches, a gardener, as a rule, cannot transmit this peculiarity by sexual, but only by non-sexual propagation. The twigs cut off such a weeping tree and planted as cuttings or slips, afterwards produce trees having likewise hanging branches, as, for example, the weeping willows and beeches. Seedlings, on the other hand, which have been reared out of the seed of such a weeping tree, generally have the original stiff and upright form of branches possessed by their ancestors. The same may be observed in a very striking manner in the so-called "copper-coloured trees," that is, varieties of trees which are characterized by a red or reddish brown colour of the leaves. Offshoots from such copper-coloured trees (for example, the copper beech), which have been propagated by cuttings in a non-sexual manner, show the peculiar colour and nature of the leaves which distinguished the parental individual, while others reared from seeds of such a copper-coloured tree return to the green-coloured condition of leaf.

This difference in inheritance will seem very natural when we consider that the material connection between the producing and produced individuals is much closer and lasts much longer in non-sexual than in sexual propagation. The special tendency of the molecular motion of life can therefore fix itself much longer and more thoroughly in the filial organism, and be more strictly transmitted by non-sexual than by sexual propagation. All these phenomena, considered in connection, clearly prove that the transmission of bodily and mental peculiarities is a purely material and mechanical process. By propagation a greater or lesser quantity of albuminous particles, and together with them the individual form of motion inherent in these molecules of protoplasm, are transmitted from the parental organism to the offspring. As this form of motion remains continuous, the more delicate peculiarities inherent in the parental organism must sooner or later reappear in the filial organism.

CHAPTER IX.

LAWS OF TRANSMISSION BY INHERITANCE. ADAPTATION AND NUTRITION.

Distinction between Conservative and Progressive Transmission by Inheritance.—Laws of Conservative Transmission: Transmission of Inherited Characters.—Uninterrupted or Continuous Transmission.—Interrupted or Latent Transmission.—Degeneracy.—Sexual Transmission.—Secondary Sexual Characters.—Mixed or Amphigonous Transmission.—Hybrids.—Abridged or Simplified Transmission.—Laws of Progressive Inheritance: Transmission of Acquired Characters.—Adapted or Acquired Transmission.—Fixed or Established Transmission.—Homochronous Transmission (Identity in Epoch).—Homotopic Transmission (Identity in Part).—Adaptation and Mutability.—Connection between Adaptation and Nutrition.—Distinction between Indirect and Direct Adaptation.

In the last chapter we considered Transmission by Inheritance, one of the two universal vital activities of organisms, Adaptation and Inheritance, which by their interaction produce the different species of organisms, and we have endeavoured to trace this very mysterious vital activity to a more general physiological function of organisms, namely, to Propagation. This latter in its turn, like other vital phenomena of animals and plants, depends on physical and chemical relations. It is true they appear at times exceedingly complicated, but can nevertheless in reality be traced to simple mechanical causes—that is, to the relations of attraction and repulsion in the particles or molecules—in fact, to the motional phenomena of matter.

Now, before we turn our attention to the second function, the phenomenon of Adaptation or Mutability, which counteracts the Transmission by Inheritance, it seems appropriate first to cast one more glance at the various manifestations of Heredity, which we may perhaps even now denominate the *"laws of transmission by inheritance."* Unfortunately, up to the present time very little has been done for this most important subject, either in zoology or in botany, and almost all we know of the different laws of inheritance is confined to the experiences of gardeners and farmers. It is not therefore to be wondered at, that on the whole these exceedingly interesting and important phenomena have not been investigated with desirable scientific accuracy, or reduced to the form of scientific laws. Accordingly, what I shall relate of the different laws of transmission are only some preliminary fragments taken out of the infinitely rich store which lies open to our inquiry.

We may first divide all the different phenomena of inheritance into two groups, which we may distinguish as the transmission of *inherited* characters, and the transmission of *acquired* characters; and we may call the former the *conservative* transmission, and the latter the *progressive* transmission by inheritance. This distinction depends upon the exceedingly important fact that the individuals of every species of animals and plants can transmit to their descendants, not only those qualities which they themselves have inherited from their ancestors, but also the peculiar, individual qualities which they have acquired during their own life. The latter are transmitted by progressive, the former by conservative inheritance. We have now first to examine the phenomena of *conservative inheritance* that is, the transmission of such qualities as the organism has already received from its parents or ancestors. (Gen. Morph. ii. 180.)

Among the phenomena of conservative inheritance we are first struck by that which is its most general law, and which we may term the *law of uninterrupted or continuous transmission*. It is so universal among the higher animals and plants, that the uninitiated might overestimate its action and consider it as the only normal law of transmission by inheritance. This law simply consists in the fact that among most species of animals and plants, every generation is, on the whole, like the preceding—that the parents are as like the grandparents as they are like the children. "Like produces like," as is commonly said, but more accurately "similar things produce similar things." For, in reality, the descendants of every organism are never absolutely equal in all points, but only similar in a greater or less degree. This law is so generally known, that I need not give any examples of it.

The *law of interrupted or latent transmission* by inheritance, which might also be termed alternating transmission, is in a measure opposed to the preceding law. This important law appears principally active among many lower animals and plants, and manifests itself in contrast to the former in the fact that the offspring are not like their parents, but very dissimilar, and that only the third or a later generation becomes similar to the first. The grandchildren are like the grandparents, but quite unlike the parents. This is a remarkable phenomenon, and, as is well known, occurs also very frequently, though in a less degree, in human families. Every one of my readers doubtless knows some members of a family who, in this or that peculiarity, much more resemble the grandfather or grandmother than the father or mother. Sometimes it lies in bodily peculiarities, for example, features of face, colour of hair, size of body—sometimes in mental qualities, for example, temperament, energy, understanding—which are transmitted in this manner. This fact may be observed in domestic animals as well as in the case of man. Among the domestic animals most liable to vary—as the dog, horse, and ox—breeders very frequently find that the product by breeding resembles the grandparents far more than it does its own parental organism. If we express this general law and the succession of generations by the letters of the alphabet, then A = C = E, whilst B = D = F, and so on.

This very remarkable fact appears in a more striking way in the lower animals and plants than in the higher, and especially in the well-known phenomenon of *alternation of generations* (metagenesis). Here we very frequently find—for example, among the Planarian worms, sea-squirts or Tunicates, Zoophytes, and also among ferns and mosses—that the organic individual in the first place produces, by propagation, a form completely different from the parental form, and that only the descendants of this generation, again, become like the first. This regular change of generation was discovered by the poet Chamisso, on his voyage round the world in 1819, among the *Salpae*, cylindrical tunicates, transparent like glass, which float on the surface of the sea. Here the larger generation, the individuals of which live isolated and possess an eye of the form of a horse-shoe, produce in a non-sexual manner (by the formation of buds) a completely different and smaller generation. The individuals of this second smaller generation live united in chains and possess a cone-shaped eye. Every individual of such a chain produces, in a sexual manner (hermaphrodite) again, a non-sexual solitary form of the first and larger generation. Among the Salpae, therefore, it is always the first, third, and fifth generation, and in like manner the second, fourth, and sixth generations, that are entirely like one another. However, it is not always only one, but in other cases a number of generations, which are thus leapt over; so that the first generation resembles the fourth and seventh, the second resembles the fifth and eighth, the third resembles the sixth and ninth, and so on. Three different generations alternate with one another; for example, among the neat *little sea-buoys* (*Doliolum*), small tunicates closely related to the Salpae. In this case it is A = D = G, further, B = E = H, and C = F = I. Among the plant-lice (Aphides), each sexual generation is followed by a succession of from eight to ten or twelve non-sexual generations, which are like one another, but differ from the sexual generations. Then, again, a sexual generation reappears like the one long before vanished.

If we further follow this remarkable law of latent or interrupted inheritance, and take into consideration all the phenomena appertaining to it, we may comprise under it also the well-known phenomena of *reversion*. By the term "reversion" or "atavism" we understand the remarkable fact known to all breeders of animals, that occasionally single and individual animals assume a form which has not existed for many generations, but belongs to a generation which has long since disappeared. One of the most remarkable instances of this kind is the fact that in some horses there sometimes appear singular dark stripes, similar to those of the zebra, quagga, and other wild species of African horses. Domestic horses of the most different races and of all colours sometimes show such dark stripes; for example, a stripe along the back, a stripe across the shoulders, and the like. The sudden appearance of these stripes can only be explained by the supposition that it is the effect of a latent transmission, a relapse into the ancient original form, which has long since vanished, and was once common to all species of horses; the original form, undoubtedly, was originally striped like the zebras, quaggas, etc. In like manner, certain qualities in other domestic animals sometimes appear quite suddenly, which once marked their wild ancestors, now long since extinct. In plants, also, such a relapse can be observed very frequently. All my readers probably know the wild yellow toad-flax (*Linaria vulgaris*), a plant very common in our fields and hedges. Its dragon-mouthed yellow flower contains two long and two short stamens. But sometimes there appears a single blossom (*Peloria*) which is funnel-shaped, and quite regularly composed of five individual and equal sections, with five corresponding stamens. This Peloria can only be explained as a relapse into the long since extinct and very ancient common form of all those plants which, like the toad-flax, possess dragon-mouthed, two-lipped flowers, with two long and two short stamens. The original form, like the Peloria, possessed a regular five-spurred blossom, with five equal stamens, which only later and by degrees have become unequal (compare p. 17 [10]). All such relapses are to be brought under the law of interrupted or latent transmission, although the number of intervening generations may be enormous.

When cultivated plants or domestic animals become wild, when they are withdrawn from the conditions of cultivated life, they experience changes which appear not only as adaptations to their new mode of life, but partially also as relapses into the ancient original form out of which the cultivated forms have been developed. Thus the different kinds of cabbage, which are exceedingly different in form, may be led back to the original form, by allowing them to grow wild. In like manner, dogs, horses, heifers, etc., when growing wild, often revert more or less to a long extinct generation. An immensely long succession of generations may pass away before this power of latent transmission becomes extinguished.

A third law of conservative transmission may be called the *law of sexual transmission*, according to which each sex transmits to the descendants of the same sex peculiarities which are not inherited by the descendants of the other sex. The so-called secondary sexual characters, which in many respects are of extraordinary interest, everywhere furnish numerous examples of this law. Subordinate or secondary sexual characters are those peculiarities of one of the two sexes which are not directly connected with the sexual organs themselves; such characters, which exclusively belong to the male sex, are, for example, the antlers of the stag, the mane of the lion, and the spur of the cock. The human beard, an ornament commonly denied to the female sex, belongs to the same class. Similar characteristics by which the female sex is alone distinguished are, for example, the developed breasts, with the lactatory glands of female mammals and the pouch of the female opossum. The bodily size, also, and complexion, differs in female animals of many species from that of the male. All these secondary sexual qualities, like the sexual organs themselves, are transmitted by the male organism only to the male, not to the female, and *vice versa*. Contrary facts are rare exceptions to the rule.

A fourth law of transmission, which has here to be mentioned, in a certain sense contradicts the last, and limits it, viz., the *law of mixed or mutual* (amphigonous) *transmission*. This law tells us that every organic individual produced in a sexual way receives qualities from both parents, from the father as well as from the mother. This fact, that personal qualities of each of the two sexes are transmitted to both male and female descendants, is very important, Goethe mentions it of himself, in the beautiful lines—

"Von Vater hab ich die Statur, des Lebens ernstes Führen
 Von Mütterchen die Frohnatur und Lust zu fabuliren."

"From my father I have my stature and the serious tenour of my life,
 From my mother a joyous nature and a turn for poetizing."

This phenomenon, I suppose, is so well-known to all, that I need not here enter upon it. It is according to the different portions of their character which father and mother transmit to their children, that the individual differences among brothers and sisters are chiefly determined.

The very important and interesting phenomenon of *hybridism* also belongs to this law of mixed or amphigonous transmission. It alone, when rightly estimated, is quite sufficient to refute the prevailing dogma of the constancy of species. Plants, as well as animals, belonging to quite different species, may sexually mingle with one another and produce descendants which in many cases can again propagate themselves, and that indeed either (more frequently) by mingling with one of the two parental species, or (more rarely) by pure in-breeding, hybrid mixing with hybrid. The latter is well established, for example, in the hybrids of hares and rabbits (*Lepus darwinii*, p. 147 [83]). The hybrids of a horse and a donkey, two different species of the same genus (*Equus*), are well known. These hybrids differ according as the father or the mother belongs to the one or the other species— the horse or the donkey. The mule produced by a mare and a he-donkey has qualities quite different from those of the jinny (*Hinnus*), the hybrid of a horse and she-donkey. In both cases the hybrid produced by the crossing of two different species is a mixed form, which receives qualities from both parents; but the qualities of the hybrid are different, according to the form of the crossing. In like manner, mulattoes produced by a European and a negress show a different mixture of characters from the hybrids produced by a negro with a European female. In these phenomena of hybrid-breeding, as well as in the other laws of transmission previously mentioned, we are as yet unable to show the acting causes in detail; but no naturalist doubts the fact that the causes are in all cases purely mechanical and dependent upon the nature of organic matter itself. If we possessed more delicate means of investigation than our rude organs of sense and auxilliary instruments, we should be able to discover those causes, and to trace them to the chemical and physical properties of matter.

Among the phenomena of conservative transmission, we must now mention, as the fifth law, the *law of abridged or simplified transmission*. This law is very important in regard to embryology or ontogeny, that is in regard to the history of the development of organic individuals. *Ontogeny*, or the history of the development of individuals, as I have already mentioned in the first chapter (p. 10 [6]), and as I subsequently shall explain more minutely, is nothing but a short and quick repetition of *Phylogeny* dependent on the laws of transmission and adaptation—that is, a repetition of the palaeontological history of development of the whole organic tribe, or phylum, to which the organism belongs. If, for example, we follow the individual development of a man, an ape, or any other higher mammal within the maternal body from the egg, we find that the foetus or embryo arising out of the egg passes through a series of very different forms, which on the whole agrees with, or at least runs parallel to, a series of forms which is presented to us by the historical chain of ancestors of the higher mammals. Among these ancestors we may mention certain fishes, amphibians, marsupials, etc. But the parallelism or agreement of these two series of development is never quite complete; on the contrary, in ontogeny there are always gaps and leaps which indicate the omission of certain stages belonging to the phylogeny. Fritz Müller, in his excellent work, "Für Darwin,"[16] has clearly shown in the case of the Crustacea, or crabs, that "the historical record preserved in the individual history of development is gradually obscured, in proportion as development takes a more and more direct route from the egg to the complete animal." This process of obscuring and shortening is determined by the law of abridged transmission, and I mention it here specially because it is of great importance for the understanding of embryology, and because it explains the fact, at first so strange, that the whole series of forms which our ancestors have passed through in their gradual development are no longer visible in the series of forms of our own individual development from the egg.

Opposed to the laws of the conservative transmission, hitherto discussed, are the phenomena of the transmission of the second series, that is, the *laws of progressive transmission by inheritance*. As already mentioned, they depend upon the fact that the organism transmits to its descendants not only those qualities which it has inherited from its own ancestors, but also a number of those individual qualities which it has acquired during its own lifetime. Adaptation is here seen to be connected with transmission by inheritance (Gen. Morph. ii. 186).

At the head of these important phenomena of progressive transmission, we may mention the *law of adapted or acquired transmission*. In reality it asserts nothing more than what I have said above, that in certain circumstances the organism is capable of transmitting to its descendants all the qualities which it has acquired during its own life by adaptation. This phenomenon, of course, shows itself most distinctly when the newly acquired peculiarity produces any considerable change in the inherited form. This is the case in the examples I mentioned in the preceding chapter as to transmission in general, in the case of the men with six fingers and toes, the porcupine men, copper beeches, weeping willows, etc. The transmission of acquired diseases, such as consumption, madness, and albinism, likewise form very striking examples. Albinoes are those individuals who are distinguished by the absence of colouring matter, or pigments, in the skin. They are of frequent occurrence among men, animals, and plants. In the case of animals of a definite dark colour, individuals are not unfrequently born which are entirely without colour, and in animals possessing eyes, this absence of pigment extends even to the eyes, so that the iris of the eye, which is commonly of a bright or intense colour, is colourless, but appears red, on account of the blood-vessels being seen through it. Among many animals, such as rabbits and mice, albinoes with white fur and red eyes are so much liked that they are propagated in great numbers as a special race. This would be impossible were it not for the law of the transmission of adaptations.

Which of the changes acquired by an organism are transmitted to its descendants, and which are not, cannot be determined *à priori*, and we are unfortunately not acquainted with the definite conditions under which the transmission takes place. We only know in a general way that certain acquired qualities are much more easily transmitted than others, for example, more easily than the mutilations caused by accidents. These latter are generally not transmitted by inheritance, otherwise the descendants of men who have lost their arms or legs would be born without the corresponding arm or leg; but here, also, exceptions occur, and a race of dogs without tails has been produced by consistently cutting off the tails of both sexes of the dog during several generations. A few years ago a case occurred on an estate near Jena, in which by a careless slamming of a stable door the tail of a bull was wrenched off, and the calves begotten by this bull were all born without a tail. This is certainly an exception; but it is very important to note the fact, that under certain unknown conditions such violent changes are transmitted in the same manner as many diseases.

In very many cases the change which is transmitted and preserved by adapted transmission is constitutional or inborn, as in the case of albinism mentioned before. The change then depends upon that form of adaptation which we call the indirect or potential. A very striking instance is furnished by the hornless cattle of Paraguay, in South America. A special race of oxen is there bred which is entirely without horns. It is descended from a single bull, which was born in 1770 of an ordinary pair of parents, and the absence of horns was the result of some unknown cause. All the descendants of this bull produced with a horned cow were entirely without horns. This quality was found advantageous, and by propagating the hornless cattle among one another, a hornless race was obtained, which at present has almost entirely supplanted the horned cattle in Paraguay. The case of the otter-sheep of North America forms a similar example. In the year 1791 a farmer, by name Seth Wright, lived in Massachusetts, in North America; in his normally formed flock of sheep a lamb was suddenly born with a surprisingly long body and very short and crooked legs. It was therefore unable to take any great leaps, and especially unable to leap across a hedge into a neighbour's garden—a quality which seemed advantageous to the owner, as the territories were divided by hedges. It therefore occurred to him to transmit this quality to other sheep, and by crossing this ram with normally shaped ewes, he produced a whole race of sheep, all of which had the qualities of the father, short and crooked legs and a long body. None of them could leap across the hedges, and they therefore were much liked and propagated in Massachusetts.

A second law, which likewise belongs to the series of progressive transmissions, may be called the *law of established or habitual transmission*. It manifests itself in this, that qualities acquired by an organism during its individual life are the more certainly transmitted to its descendants the longer the causes of that change have been in action, and that this change becomes the more certainly the property of all subsequent generations the longer the cause of change acts upon these latter also. The quality newly acquired by adaptation or mutation must be established or constituted to a certain degree before we can calculate with any probability that it will be transmitted at all to the descendants. In this respect transmission resembles adaptation. The longer a newly acquired quality has been transmitted by inheritance, the more certainly will it be preserved in future generations. If, therefore, for example, a gardener by methodical treatment has produced a new kind of apple, he may calculate with the greater certainty upon preserving the desired peculiarity of this sort the longer he has transmitted the same by inheritance. The same is clearly shown in the transmission of diseases. The longer consumption or madness has been hereditary in a family the deeper is the root of the evil, and the more probable it is that all succeeding generations will suffer from it.

We may conclude the consideration of the phenomena of inheritance with the two very important laws of *homotopic* and *contemporaneous transmission by inheritance*. We understand by them the fact that changes acquired by an organism during its life, and transmitted to its descendants, appear in the same part of the body in which the parental organism was first affected by them, and that they also appear in the offspring at the same age as that at which they did so in the parent.

The law of contemporaneous or homochronous transmission, which Darwin calls the law of "transmission in corresponding periods of life," can be shown very clearly in the transmission of diseases, especially of such as are recognized as very destructive, on account of their hereditary character. They generally appear in the organism of the child at the time corresponding with that in which the parental organism contracted the disease. Hereditary diseases of the lungs, liver, teeth, brain, skin, etc., usually appear in the descendants at the same period, or a little earlier than they showed themselves in the parental organism, or were contracted by it. The calf gets its horns at the same period of life as its parents did. In like manner the young stag receives its antlers at the same period of life in which they appeared in its father or grandfather. In every one of the different sorts of vine the grapes ripen at the same time as they did in the case of their progenitors. It is well known that the time of ripening varies greatly in the different sorts; but as all are descended from a single species, this variation has been acquired by the progenitors of the several sorts, and has then been transmitted by inheritance.

The *law of homotopic transmission*, which is most closely connected with the last mentioned law, and which might be called the law of transmission in corresponding parts of the body, may also be very distinctly recognized in pathological cases of inheritance. Large moles, for example, or accumulations of pigment in several parts of the skin, tumours also, often appear during many generations, not only at the same period of life, but also in the same part of the skin. Excessive development of fat in certain parts of the body is likewise transmitted by inheritance. Above all, it is to be noted that numerous examples of this, as well as of the preceding law, may be found everywhere in the study of embryology. Both the *law of homochronous and homotopic transmission are fundamental laws of embryology, or ontogeny.* For these laws explain the remarkable fact that the different successive forms of individual development in all generations of one and the same species always appear in the same order of succession, and that the variations of the body always take place in the same parts. This apparently simple and self-evident phenomenon is nevertheless exceedingly wonderful and curious; we cannot explain its real causes, but may confidently assert that they are due to the direct transmission of the organic matter from the parental organism to that of the offspring, as we have seen above in the case of the process of transmission in general, by a consideration of the details of the various modes of reproduction.

Having thus, then, considered the most important laws of Inheritance, we now turn to the second series of phenomena bearing on natural selection, viz., to those of Adaptation or Variation. These phenomena, taken as a whole, stand in a certain opposition to the phenomena of Inheritance, and the difficulty which arises in examining them consists mainly in the two sets of phenomena being so completely intercrossed and interwoven. We are but seldom able to say with certainty—of the variations of form which occur before our eyes—how much is owing to Inheritance, and how much to

Adaptation. All characters of form, by which organisms are distinguished, are caused *either* by Inheritance or by Adaptation; but as both functions are continually interacting with each other, it is extremely difficult for the systematic inquirer to recognize the share belonging to each of the two functions in the special structure of individual forms. This is, at present, all the more difficult, because we are as yet scarcely aware of the immense importance of this fact, and because most naturalists have neglected the theory of Adaptation, as well as that of Inheritance. The laws of Inheritance, which we have just discussed, as well as the laws of Adaptation, which we shall consider directly, in reality form only a small portion of the phenomena existing in this domain, but which have not as yet been investigated; and since every one of these laws can interact with every other, it is clear that there is an infinite complication of physiological actions, which are at work in the construction of organisms.

But now, as to the phenomenon of variation or adaptation in general, we must, as in the case of inheritance, view it as a quite universal, physiological fundamental quality of all organisms, without exception—as a manifestation of life which cannot be separated from the idea of organism. Strictly speaking, we must here also, as in the case of inheritance, distinguish between Adaptation itself and Adaptability. By Adaptation (Adaptio), or Variation (Variatio), we understand the fact that the organism, in consequence of influences of the surrounding outer world, assumes certain new peculiarities in its vital activity, composition, and form which it has not inherited from its parents; these acquired individual qualities are opposed to those which have been inherited, or, in other words, those which have been transmitted to it from its parents or ancestors. On the other hand, we call Adaptability (Adaptabilitas), or Variability (Variabilitas), the capability inherent in all organisms to acquire such new qualities under the influence of the outer world. (Gen. Morph. ii. 191.)

The undeniable fact of organic adaptation or variation is universally known, and can be observed at every moment in thousands of phenomena surrounding us. But just because the phenomena of variation by external influences appear so self-evident, they have hitherto undergone scarcely any accurate scientific investigation. To them belong all the phenomena which we look upon as the results of contracting and giving up habits, of practice and giving up practices, or as the results of training, of education, of acclimatization, of gymnastics, etc. Many permanent variations brought about by causes producing disease, that is to say, many diseases, are nothing but dangerous adaptations of the organism to injurious conditions of life. In the case of cultivated plants and domestic animals, variation is so striking and powerful that the breeder of animals and the gardener found their whole mode of proceeding upon it, or rather upon the interaction between these phenomena and those of Inheritance. It is also well known to every one that animals and plants, in their wild state, are subject to variation. Every systematic treatise on a group of animals or plants, if it were to be quite complete and exhaustive, ought to mention in every individual species the number of variations which differ more or less from the prevailing or typical form of the species. Indeed, in every careful systematic special treatise one finds, in the case of most species, mention of a number of such variations, which are described sometimes as individual deviations, and sometimes as so-called races, varieties, degenerate species, or subordinate species, and which often differ exceedingly from the original species, solely in consequence of the adaptation of the organism to the external conditions of life.

If we now endeavour to fathom the general causes of these phenomena of Adaptation, we arrive at the conclusion that in reality they are as simple as the causes of the phenomena of Inheritance. We have shown that the nature of the process of propagation furnishes the real explanation of the facts of Transmission by Inheritance, that is, the transmission of parental matter to the body of the offspring; and in like manner we can show that the physiological function of *nutrition*, or *change of substance*, affords a general explanation of Adaptation or Variation. When I here point to "nutrition" as the fundamental cause of variation and adaptation, I take this word in its widest sense, and I understand by it the whole of the material changes which the organism undergoes in all its parts through the influences of the surrounding outer world. Nutrition thus comprises not only the reception of actual nutritive substances and the influence of different kinds of food, but also, for example, the action upon the organism of water and of the atmosphere, the influence of sunlight, of temperature, and of all those meteorological phenomena which are implied in the term "climate." The indirect and direct influence of the nature of the soil and of the dwelling-place also belong to it; and further, the extremely important and varied influence which is exercised upon every animal and every plant by the surrounding organisms, friends and neighbours, enemies and robbers, parasites, etc. All these and many other very important influences, all of which more or less modify the organism in its material composition, must be taken into consideration in studying the change of substance which goes on in living things. Adaptation, accordingly, is the consequence of all those material variations which are produced in the change of substance of the organism by the external conditions of existence, or by the influences of the surrounding external world.

How very much every organism is dependent upon the whole of its external surroundings, and changed by their alteration, is, in a general way, well known to every one. Only think how much the human power of action is dependent upon the temperature of the air, or how much the disposition of our minds depends upon the colour of the sky. Accordingly as the sky is cloudless and sunny, or covered with large heavy clouds, our state of mind is cheerful or dull. How differently do we feel and think in a forest during a stormy winter night and during a bright summer day! All the different moods of our soul depend upon purely material changes of our brain, upon movements of molecular plasma, which are started through the medium of the senses by the different influences of light, warmth, moisture, etc. "We are a plaything to every pressure of the air." No less important and deeply influential are the effects produced upon our mind and body by the different quality and quantity of food. Our mental activity, the activity of our understanding and of our imagination, is quite different accordingly as we have taken tea or coffee, wine or beer, before or during our work. Our moods, wishes, and feelings are quite different when we are hungry and when we are satisfied. The national character of Englishmen and Gauchos, in South America, who live principally on meat and food rich in nitrogen, is wholly different from that of the Irish, feeding on potatoes, and that of the Chinese, living on rice, both of whom take food deficient in nitrogen. The latter also form much more fat than the former. Here, as everywhere, the variations of the mind go hand in hand with the corresponding transformations of the body; both are produced by purely material causes. But all other organisms, in the same way as man, are varied and changed by the different influences of nutrition. It is well known that we can change in an arbitrary way the form, size, colour, etc., of our cultivated plants and domestic animals, by change of food; that, for example, we can take from or give to a plant definite qualities, accordingly as we expose it to a greater or less degree of sunlight and moisture. As these phenomena are generally widely known, and as we shall proceed presently to the consideration of the different laws of adaptation, we will not dwell here any longer on the general facts of variation.

As the different laws of transmission may be naturally divided into the two series of conservative and progressive transmission, so we may also distinguish between two series of the laws of adaptation, first, the series of laws of *indirect*, and secondly, the series of laws of *direct* adaptation. The latter may also be called the laws of actual, and the former the laws of potential, adaptation.

The first series, comprising the phenomena of *indirect* (potential) adaptation, has, on the whole, hitherto been little attended to, and Darwin has the merit of having directed special attention to this series of changes. It is somewhat difficult to place this subject clearly before the reader; I will endeavour to make it clear hereafter by examples. Speaking quite generally, indirect or potential adaptation consists in the fact that certain changes in the organism, effected by the influence of nutrition (in its widest sense) and of the external conditions of existence in general, show themselves not in the individual form of the respective organism, but in that of its descendants. Thus, especially in organisms propagating themselves in a sexual way, the reproductive system, or sexual apparatus, is often influenced by external causes (which little affect the rest of the organism), to such a degree that its descendants show a complete alteration of form. This can be seen very strikingly in artificially produced monstrosities. Monstrosities can be produced by subjecting the parental organism to certain extraordinary conditions of life, and, curiously enough, such an extraordinary condition of life does not produce a change of the organism itself, but a change in its descendants. This cannot be called transmission by inheritance, because it is not a quality existing in the parental organism that is transmitted by inheritance. It is, on the contrary, a change affecting the parental organism, but not perceptible in it, that appears in the peculiar formation of its descendants. It is only the impulse to this new formation which is transmitted in propagation through the egg of the mother or the sperm of the father. The new formation exists in the parental organism only as a possibility (potential); in the descendants it becomes a reality (actual).

As this very important and very general phenomenon had hitherto been entirely neglected, people were inclined to consider all the visible variations and transformations of organic forms as phenomena of adaptation of the second series, that is, as phenomena of *direct* or actual adaptation. The essence of this latter kind of adaptation consists in the fact that the change affecting the organism (through nutrition, etc.) shows itself immediately by some transformation, and does not only make itself apparent in the descendants. To this class belong all the well-known phenomena in which we can directly trace the transforming influence of climate, food, education, training, etc., in their effects upon the individual itself.

We have seen how the two series of phenomena of progressive and conservative transmission, in spite of their difference in principle, in many ways interfere with and modify each other, and in many ways co-operate with and cross each other. The same is the case, in a still higher degree, in the two series of phenomena of indirect and direct adaptation, which are opposed to each other and yet closely connected. Some naturalists, especially Darwin and Carl Vogt, ascribe to the indirect or potential adaptation by far the more important and almost exclusive influence. But the majority of naturalists have hitherto been inclined to take the opposite view, and to attribute the principal influence to direct or actual adaptation. I consider this controversy, in the mean while, as almost useless. It is but seldom that we are in a condition, in any individual case of variation, to judge how much of it belongs to direct and how much to indirect adaptation. We are, on the whole, still too little acquainted with these exceedingly important and intricate relations, and can only assert, in a general way, that the transformation of organic forms is to be ascribed *either* to direct adaptation alone, *or* to indirect adaptation alone, or lastly, to the co-operation of both direct *and* indirect adaptation.

CHAPTER X.

LAWS OF ADAPTATION.

Laws of Indirect or Potential Adaptation.—Individual Adaptation.—Monstrous or Sudden Adaptation.—Sexual Adaptation.—Laws of Direct or Actual Adaptation.—Universal Adaptation.—Cumulative Adaptation.—Cumulative Influence of External Conditions of Existence and Cumulative Counter-Influence of the Organism.—Free Will.—Use and Non-use of Organs.—Practice and Habit.—Correlative Adaptation.—Correlation of Development.—Correlation of Organs.—Explanation of Indirect or Potential Adaptation by the Correlation of the Sexual Organs and of the other parts of the Body.—Divergent Adaptation.—Unlimited or Infinite Adaptation.

In the last chapter we reduced into two groups the phenomena of Adaptation or Variation, which, in connection and interaction with the phenomena of Heredity, produce all the endless variety of forms in animals and plants— first, the group of indirect or potential, and secondly, the group of direct or actual Adaptation. We shall occupy ourselves with a closer examination of the different laws which we can discover in these two groups of the phenomena of variation. Let us first take into consideration the remarkable and very important, although hitherto much neglected, phenomena of indirect variation.

Indirect or potential adaptation manifests itself, it will be remembered, in the striking and exceedingly important fact that organic individuals experience transformations and assume forms in consequence of changes of nutrition which have not operated on them themselves, but upon their parental organism. The transforming influence of the external conditions of existence, of climate, of nutrition, etc., shows its effects here not directly in the transformation of the organism itself, but indirectly in that of its descendants. (Gen. Morph. ii. 202.)

As the principal and most universal of the laws of indirect variation must be mentioned *the law of individual adaptation*, or the important proposition that all organic individuals from the commencement of their individual existence are unequal, although often very much alike. As a proof of this proposition, I may at once point to the fact, that in the human race in general all brothers and sisters, all children of the same parents, are unequal from their birth. No one will venture to assert that two children at their birth are perfectly alike; that the size of the individual parts of their bodies, the number of hairs on their heads, the number of cells composing their outer skins or epidermis, the number of blood-cells are the same in both children, or that both children have come into the world with the same abilities or talents. But what more specially proves this law of individual difference, is the fact that in the case of those animals which produce several young ones at a time,—for instance, dogs and cats,—all the young of each birth differ from one another more or less strikingly in size and colour of the individual parts of the body, or in strength, etc. Now this law is universal. All organic individuals from their beginning are distinguished by certain, though often extremely minute, differences, and the cause of these individual differences, though in detail usually utterly unknown to us, depends partly or entirely on certain influences which the organs of propagation in the parental organism have undergone.

A second law of indirect adaptation, which we shall call *the law of monstrous or sudden adaptation*, is of less importance and less general than the law of individual adaptation. Here the divergences of the child-organism from the parental form are so striking that, as a rule, we may designate them as monstrosities. In many cases they are produced, as has been proved by experiments, by the parental organism having been subject to a certain treatment, and placed under peculiar conditions of nutrition; for example, when air and light are withdrawn from it, or when other influences powerfully acting upon its nutrition are changed in a certain way. The new condition of existence causes a strong and striking modification of form, not directly of the organism itself, but only of that of its descendants. The mode of this influence in detail we cannot discover, and we can only in a very general way detect a causal connection between the abnormal formation of the child and a certain change in the conditions of existence of its parents exerting a special influence upon the organs of propagation in the latter. The previously mentioned phenomenon of albinism probably belongs to this group of abnormal or sudden variations, also the individual cases of human beings with six fingers and toes, the case of the hornless cattle, as well as those of sheep and goats with four or six horns. The abnormal deviation in all these cases probably owes its origin to a cause which at first only affected the reproductive system of the parental organism, the egg of the mother or the sperm of the father.

A third curious manifestation of indirect adaptation may be termed *the law of sexual adaptation*. Under this name we indicate the remarkable fact that certain influences, which act upon the male organs of propagation only, affect the structure of the male descendants, and in like manner other influences, which act upon the female organs of propagation only, manifest their effect only in the change of structure of the female descendants. This remarkable phenomenon is still very obscure, and has not as yet been investigated, but is probably of great importance in regard to the origin of "secondary sexual characteristics," to which we have already made allusion.

All the phenomena of sexual, monstrous, and individual adaptation, which we may comprise under the name of the laws of *indirect or potential adaptation*, are as yet very little known to us in their real nature and in their deeper causal connection. Only this much we can at present maintain with certainty, that numerous and important transformations in organic forms owe their existence to this process. Many and striking variations of form solely depend on causes which at first only affect the nutrition of the parental organism, and specially its organs of propagation. Evidently the relations in which the sexual organs stand to other parts of the body are of the greatest importance. We shall have more to say of these presently, when we speak of the law of correlative adaptation. How powerfully the variations in the conditions of life and nutrition affect the propagation of organisms is rendered obvious by the remarkable fact that numerous wild animals which we keep in our zoological gardens, and exotic plants which are grown in our botanical gardens, are no longer able to reproduce themselves. This is the case, for example, with most birds of prey, parrots, and monkeys. The elephant, also, and the animals of prey of the bear genus, in captivity hardly ever produce young ones. In like manner many plants in a cultivated state become sterile. The two sexes may indeed unite, but no fructification, or no development of the fructified germ, takes place. From this it follows with certainty that the changed mode of nutrition in the cultivated state is able completely to destroy the capability of reproduction, and therefore to exercise the greatest influence upon the sexual organs. In like manner other adaptations or variations of nutrition in the parental organism may cause, not indeed a complete want of descendants, but still important changes in their form.

Much better known than the phenomena of indirect or potential adaptation are those of *direct or actual adaptation*, to the consideration of which we now turn our attention. To them belong all those changes of organisms which are generally considered to be the results of practice, habit, training, education, etc.; also those changes of organic forms which are effected directly by the influence of nutrition, of climate, and other external conditions of existence. As has already been remarked in direct or actual adaptation, the transforming influence of the external cause affects the form of the organism itself, and does not only manifest itself in that of the descendants. (Gen. Morph. ii. 207.)

We may place *the law of universal adaptation* at the head of the different laws of direct or actual adaptation, because it is the chief and most comprehensive among them. It may be briefly explained in the following proposition: "All organic individuals become unequal to one another in the course of their life by adaptation to different conditions of life, although the individuals of one and the same species remain mostly very much alike." A certain inequality of organic individuals, as we have seen, was already to be assumed in virtue of the law of individual (indirect) adaptation. But, beyond this, the original inequality of individuals is afterwards increased by the fact that every individual, during its own independent life, subjects and adapts itself to its own peculiar conditions of existence. All different individuals of every species, however like they may be in their first stages of life, become in the further course of their existence less like to one another. They deviate from one another in more or less important peculiarities, and this is a natural consequence of the different conditions under which the individuals live. There are no two single individuals of any species which can complete their life under exactly the same external circumstances. The vital conditions of nutrition, of moisture, air, light; further, the vital conditions of society, the inter-relations with surrounding individuals of the same or other species, are different in every individual being; and this difference first affects the functions, and later changes the form of every individual organism. If the children of a human family show, even at the beginning, certain individual inequalities which we may consider as the consequence of individual (indirect) adaptation, they will appear still more different at a later period of life, when each child has passed through different experiences, and has adapted itself to different conditions of life. The original difference of the individual processes of development, evidently becomes greater the longer the life lasts and the more various the external conditions which influence the separate individuals. This may be demonstrated in the simplest manner in man, as well as in domestic animals and cultivated plants, in which the vital conditions may be arbitrarily modified. Two brothers, of whom one is brought up as a workman and the other as a priest, develop quite differently in body as well as in mind; in like manner, two dogs of one and the same birth, of which one is trained as a sporting dog and the other chained up as a watch dog. The same observation may also readily be made as to organic individuals in a natural state. If, for instance, one carefully compares all the trees in a fir or beech forest, which consists of trees of a single species, one finds that among all the hundreds or thousands of trees, there are not two individual trees completely agreeing in size of trunk and other parts, in the number of branches, leaves, etc. Everywhere we find individual inequalities which, in part at least, are merely the consequences of the different conditions of life under which the trees have developed. It is true we can never say with certainty how much of this dissimilarity in all the individuals of every species may have originally been caused by indirect individual adaptation, and how much of it acquired under the influence of direct or universal adaptation.

A second series of phenomena of direct adaptation, which we may comprise under *the law of cumulative adaptation,* is no less important and general than universal adaptation. Under this name I include a great number of very important phenomena, which are usually divided into two quite distinct groups. Naturalists, as a rule, have distinguished, first, those variations of organisms which are produced directly by the permanent influence of external conditions (by the constant action of nutrition, of climate, of surroundings, etc.), and secondly, those variations which arise from habit and practice, from accustoming themselves to definite conditions of life, and from the use and non-use of organs. The latter influences have been set forth especially by Lamarck as important causes of the change of organic forms, while the former have for a very long time been recognized as such more generally.

The sharp distinction usually made between these two groups of cumulative adaptation, and which even Darwin still maintains, disappears as soon as we reflect more accurately and deeply upon the real nature and causal foundation of these two, apparently very different, series of adaptations. We then arrive at the conviction that in both cases there are always two different active causes to be dealt with: on the one hand the *external influence* or *action* of adaptative conditions of life, and on the other hand the *internal reaction of the organism* which subjects and adapts itself to that condition of life. If cumulative adaptation is considered from the first point of view alone, and the transforming actions of the permanent external conditions of life are traced to those conditions solely, then the principal stress is laid unduly upon the external factor, and the necessary internal reaction of the organism is not taken into proper consideration. If, on the other hand, cumulative adaptation is unjustly regarded solely in relation to its second factor, and the transforming action of the organism itself, its reaction against the external influences, its change by practice, habit, use, or non-use of organs, is put into the foreground, then we forget that this reaction is first called into play by the action of external conditions of existence. Hence it seems that the distinction made between these two groups lies only in the different manner of viewing them, and I believe that they can, with full justice, be considered as one. The most essential fact in these phenomena of cumulative adaptation is that the change of the organism which manifests itself first in the functions, and at a later period in the form, is the result either of long enduring, or of often repeated, influences of an external cause. The smallest cause, by cumulation of its action, can attain the greatest results.

There are innumerable examples of this kind of direct adaptation. In whatever direction we may examine the life of animals and plants, we discover on all hands evident and undeniable changes of this kind. Let me first mention some of those phenomena of adaptation occasioned directly by nutrition itself. Every one knows that the domestic animals which are bred for certain purposes can be variously modified, according to the different quantity and quality of the food given to them. If a farmer in breeding sheep wishes to produce fine wool, he gives them different food from what he would give if he wished to obtain good flesh or an abundance of fat. Choice race and carriage horses receive better food than dray and cart horses. Even the bodily form of man—for example, the amount of fat—is quite different according to his nutrition. Food containing much nitrogen produces little fat, that containing little nitrogen produces a great deal of fat. People who, by means of Banting's system, at present so popular, wish to become thin eat only meat and eggs—no bread, no potatoes. The important variations that can be produced among cultivated plants, solely by changing the quantity and quality of nourishment, are well known. The same plant acquires an altogether different appearance, according as it is placed in a dry and warm place, exposed to the sunlight or placed in a cool damp spot in the shade. Many plants, if transferred to the sea shore, get in a short space of time thick, fleshy leaves, and the same plants placed in a particularly dry and hot locality get thin hairy leaves. All these variations arise directly from the cumulative influence of changed nutrition.

But it is not only the quantity and quality of the articles of nutrition which affect and powerfully change and transform the organism, but it is affected also by all the other external conditions of existence, above all by its nearest organic surroundings, the society of friendly or hostile organisms. One and the same kind of tree develops itself quite differently in an open locality, where it is free on all sides, and in a forest where it must adapt itself to its surroundings, where it is pressed on all sides by its nearest neighbours, and is forced to shoot upwards. In the former case, the branches of the tree spread widely out; in the latter, the trunk extends upwards, and the top of the tree remains small and contracted. How powerfully all these circumstances, and how powerfully the hostile or friendly influence of surrounding organisms, of parasites, etc., affect every animal and every plant, is so well known, that it appears superfluous to quote further examples. The change of form, or transformation which is thereby effected, is never solely the direct result of the external influence, but must always be traced to the corresponding reaction, and to the activity of the organism itself, which consists in contracting a habit, or practice, and in the use or non-use of organs. The fact that these latter phenomena, as a rule, have been considered distinct from the former, is owing first to the one-sided manner of viewing them already mentioned, and secondly to the wrong notion which has been formed as to the nature and the influence of the activity of the will in animals.

The activity of the will, which is the organ of habit, of practice, of the use or non-use of organs among animals, is, like every other activity of the animal soul, dependent upon material processes in the central nervous system, upon peculiar motions which emanate from the albuminous matter of the ganglion cells, and the nervous fibres connected with them. The will, as well as the other mental activities, in higher animals, in this respect is different from that of men only in quantity, not in quality. The will of the animal, as well as that of man, is never free. The widely spread dogma of the freedom of the will is, from a scientific point of view, altogether untenable. Every physiologist who scientifically investigates the activity of the will in man and animals, must of necessity arrive at the conviction that *in reality the will is never free*, but is always determined by external or internal influences. These influences are for the most part ideas which have been either formed by Adaptation or by Inheritance, and are traceable to one or other of these two physiological functions. As soon as we strictly examine the action of our own will, without the traditional prejudice about its freedom, we perceive that every apparently free action of the will is the result of previous ideas, which are based on notions inherited or otherwise acquired, and are therefore, in the end, dependent on the laws of Adaptation and Inheritance. The same also applies to the action of the will in all animals. As soon as their will is considered in connection with their mode of life, in its relation to the changes which the mode of life is subject to from external conditions, we are at once convinced that no other view is possible. Hence the changes of the will which follow the changes of nutrition, and which, in the form of practice, habit, etc., produce variations in structure, must be reckoned among the other material processes of cumulative adaptation.

Whilst an animal's will is adapting itself to changed conditions of existence by the acquisition of new habits, practices, etc., it not unfrequently effects the most remarkable transformations of the organic form. Numerous instances of this may be found everywhere in animal life. Thus, for example, many organs in domestic animals are suppressed, when in consequence of a changed mode of life they cease to act. Ducks and fowls in a wild state fly exceedingly well, but lose this facility more or less in a cultivated state. They accustom themselves to use their legs more than their wings, and in consequence the muscles and skeleton used in flying are essentially changed in their development and form. Darwin has proved this by a very careful comparative measurement and weighing of the respective parts of the skeleton in the different races of domestic ducks, which are all descended from the wild duck (*Anas boschas*). The bones of the wings in tame ducks are weaker, the bones of the legs, on the other hand, are more strongly developed than in wild ducks. In ostriches and other running birds which have become completely unaccustomed to fly, the consequence is that their wings are entirely crippled and degenerate into mere "rudimentary organs" (p. 12 [7]). In many domestic animals, especially in many races of dogs and rabbits, we find that in the cultivated state they have acquired pendulous ears. This is simply a consequence of a diminished use of the auricular muscles. In a wild state these animals have to exert their ears very much in order to discover an approaching foe, and this is accompanied by a strong development of the muscular apparatus, which keeps the outer ears in an upright position, and by which they can turn them in all directions. In a domestic state the same animals no longer require to listen so attentively, they prick up or turn their ears only a little; the auricular muscles cease to be used, gradually become weakened, and the ears hang down flabbily, or become rudimentary.

As in these cases the function, and consequently the form also, of the organ becomes degenerated through disuse, so, on the other hand, it becomes more developed by greater use. This is particularly striking if we compare the brain, and the mental activity belonging to it, in wild animals and those domestic animals which are descended from them. The dog and horse, which are so vastly improved by cultivation, show an extraordinary degree of mental development, in comparison with their wild original ancestors, and evidently the change in the bulk of the brain, which is connected with it, is mainly determined by persistent exercise. It is also well known how quickly and powerfully muscles grow and change their form by continual practice. Compare, for example, the arms and legs of a trained gymnast with those of an immovable book-worm.

How powerfully external influences affect the habits of animals and their mode of life, and in this way still further change their forms, is very strikingly shown in many cases among amphibious animals and reptiles. Our commonest indigenous snake, the ringed snake, lays eggs which require three weeks' time to develop. But when it is kept in captivity, and no sand is strewn in the cage, it does not lay its eggs, but retains them until the young ones are developed. The difference between animals producing living offspring and those laying eggs is here effaced simply by the change of the ground upon which the animal lives.

The water-salamanders, or tritons, which have been artificially made to retain their original gills, are extremely interesting in this respect. The tritons are amphibious animals, nearly akin to frogs, and possess, like the latter, in their youth external organs of respiration—gills—with which they, while living in water, breathe the air dissolved in the water. At a later date a metamorphosis takes place in tritons, as in frogs. They leave the water, lose their gills, and accustom themselves to breathe with their lungs. But if they are prevented from doing this by being kept shut up in a tank, they do not lose their gills. The gills remain, and the water salamander continues through life in that low stage of development, beyond which its lower relations, the gilled salamanders, or Sozobranchiata, never pass. The gilled salamander attains its full size, its sexual development, and reproduces itself without losing its gills.

Great interest was caused a short time ago, among zoologists, by the axolotel (*Siredon pisciformis*), a gilled salamander from Mexico, nearly related to the triton; it had already been known for a long time, and been bred on a large scale in the zoological garden in Paris. This animal possesses external gills, like the young salamander, but retains them all its life, like all other Sozobranchiata. This gilled salamander generally remains in the water, with its aquatic organs of respiration, and also propagates itself there. But in the Paris garden, unexpectedly from among hundreds of these animals, a small number crept out of the water on to the dry land, lost their gills, and changed themselves into gill-less salamanders, which are not to be distinguished from a North-American genus of tritons (*Amblystoma*), and breathe only through lungs. In this exceedingly curious case we can directly follow the great stride from water-breathing to air-breathing animals, a stride which can indeed be observed every spring in the individual history of development of frogs and salamanders. Just as every separate frog and every separate salamander transforms itself from an amphibious animal breathing through gills, at a later period into one breathing through lungs, so the whole group of frogs and salamanders have arisen from animals breathing through gills, and akin to the Siredon. The Sozobranchiata have remained up to the present day in that low stage of development. Ontogeny here explains phylogeny; the history of the development of individuals explains that of the whole group (p. 10 [6]).

To the law of accumulative adaptation there closely follows a third law of direct or actual adaptation, *the law of correlative adaptation.* According to this important law, actual adaptation not only changes those parts of the organism which are directly affected by its influence, but other parts also not directly affected by it. This is the consequence of organic solidarity, and especially of the unity of the nutrition existing among all the parts of every organism. If, for example, the hairiness of the leaves increases in a plant by its being transferred to a dry locality, then this change reacts upon the nutrition of other parts, and it may result in a shortening of the parts of the stalk, and produce a more contracted form of the whole plant. In some races of pigs and dogs—for example, in the Turkish dog—which by adaptation to a warmer climate have more or less lost their hair, the teeth also have degenerated. Whales and Endentata (armadillos), which by their curious skin-covering are removed from the other mammals, also show the greatest deviations in the formation of their teeth. Further, those races of domestic animals (oxen and pigs) which have acquired short legs have, as a rule, also a short and compact head. Among other examples, the races of pigeons which have the longest legs are also characterized by the longest beaks. The same correlation between the length of the legs and beaks is universal in the order of stilted-birds (Grallatores), in storks, cranes, snipe, etc. The correlations which thus exist between different parts of the organism are most remarkable, but their real cause is unknown to us. In general, we can of course say, the changes of nutrition affecting an individual part must necessarily react on the other parts, because the nutrition of every organism is a connected, centralized activity. But why just this or that part should exhibit this or that particular correlation is in most cases quite unknown to us. We know a great number of such correlations in nutrition; they are especially seen in those changes of animals and plants which give rise to an absence of pigment (noticed previously)—in albinoes. The want of the usual colouring matter goes hand in hand with certain changes in the formation of other parts; for example, of the muscular and osseous system, consequently of organic systems which are not at all ultimately connected with the system of the outer skin. Very frequently albinoes are more feebly developed, and consequently the whole structure of the body is more delicate and weak than in coloured animals of the same species. The organs of the senses and nervous system are in like manner curiously affected when there is this want of pigment. White cats with blue eyes are nearly always deaf. White horses are distinguished from coloured horses by their special liability to form sarkomatous tumours. In man, also, the degree of the development of pigment in the outer skin greatly influences the susceptibility of the organism for certain diseases; so that, for instance, Europeans with a dark complexion, black hair, and brown eyes become more easily acclimatized to tropical countries, and are less subject to the diseases there prevalent (inflammation of the liver, yellow fever, etc.) than Europeans of white complexion, fair hair, and blue eyes. (Compare above, p. 150 [85])

Among these correlations in the formation of different organs, those are specially remarkable which exist between the sexual organs and other parts of the body. No change of any part reacts so powerfully upon the other parts of the body as a certain treatment of the sexual organs. Farmers who wish to obtain an abundant formation of fat in pigs, sheep, etc., remove the sexual organs by cutting them out (castration), and this is indeed done to animals of both sexes. The result is an excessive development of fat. The same is done to the singers in certain religious corporations. These unfortunates are castrated in early youth, in order that they may retain their high boyish voices. In consequence of this mutilation of the genitals, the larynx remains in its youthful stage of development. The muscular tissues of the body remain at the same time weakly developed, while below the skin an abundance of fat accumulates. But this mutilation also powerfully reacts upon the development of the nervous system, the energy of the will, etc., and it is well known that human castrates, or eunuchs, as well as castrated animals, are utterly deficient in the special psychical character which distinguishes the male sex. Man is a man, both in body and soul, solely through his male generative glands.

These most important and influential correlations between the sexual organs and the other parts of the body, especially the brain, are found equally in both sexes. This might be expected even à priori, because in most animals the two kinds of organs develop themselves from the same foundation, and at the beginning are not different. In man, as in the rest of the vertebrate animals, the male and female organs in the original state of the germ are entirely the same, and the differences of the two sexes only gradually arise in the course of embryonic development (in man, in the ninth week of embryonic life), by one and the same gland developing in the female as the ovary, and in the male as the testicle. Every change of the female ovary, therefore, has a no less important reaction upon the whole female organism than every change of the testicle has upon the male organism. Virchow has expressed the importance of this correlation in his admirable essay on "Das Weib und die Zelle" ("Woman and the Cell"), in the following words:—"Woman is woman only by her sexual glands; all the peculiarities of her body and mind, of her nutrition and her nervous activity, the sweet delicacy and roundness of her limbs, the peculiar formation of the pelvis, the development of the breasts, the continuance of the high voice, that beautiful ornament of hair on her head, with the scarcely perceptible soft down on the rest of the skin—then again, the depth of feeling, the truth of her direct perceptions, her gentleness, devotion, and fidelity—in short, all the feminine qualities which we admire and honour in a true woman are but a dependence of the ovary. Take this ovary away, and the man-woman stands before us—a loathly abortion."

The same close correlation between the sexual organs and the other parts of the body occurs among plants as generally as among animals. If one wishes to obtain an abundance of fruit from a garden plant, the growth of the leaves is curtailed by cutting off some of them. If, on the other hand, an ornamental plant with a luxuriance of large and beautiful leaves is desired, then the development of the blossoms and fruit is prevented by cutting off the flower buds. In both cases one system of organs develops at the cost of the others. Thus, also, most variations in the formation of leaves in wild plants result in corresponding transformations of the generative parts or blossoms. The great importance of this "compensation of development," of this "correlation of parts," has been already set forth by Goethe, by Geoffroy St. Hilaire, and other nature-philosophers. It rests mainly upon the fact that direct or actual adaptation cannot produce an important change in a single part of the body, without at the same time affecting the whole organism.

The correlative adaptation between the reproductive organs and the other parts of the body deserves a very special consideration, because it is, above all others, likely to throw light upon the obscure and mysterious phenomena of indirect or potential adaptation, which have already been considered. For just as every change of the sexual organs powerfully reacts upon the rest of the body, so on the other hand every important change in another part of the body must necessarily more or less react on the sexual organs. This reaction, however, will only become perceptible in the formation of the offspring which arise out of the changed generative parts. It is, in fact, precisely those remarkable and imperceptible changes of the genital system (in themselves utterly insignificant changes)—changes of the eggs and the sperm—brought about by such correlations, which have the greatest influence upon the formation of the offspring, and all the phenomena of indirect or potential adaptation previously mentioned may in the end be traced to correlative adaptation.

A further series of remarkable examples of correlative adaptation is furnished by the different animals and plants which become degenerated through parasitic life or parasitism. No other change in the mode of life so much affects the shapes of organisms as the adoption of a parasitical life. Plants thereby lose their green leaves; as, for instance, our native parasitical plants, Orobanche, Lathraea, Monotropa. Animals which originally have lived freely and independently, but afterwards adopt a parasitical mode of life on other animals or plants, in the first place cease to use their organs of motion and their organs of sense. The loss of this activity is succeeded by the loss of the organs themselves, and thus we find, for example, many crabs, or crustacea, which in their youth possess a tolerably high degree of organization, viz., legs, antennae, and eyes, in old age completely degenerate, living as parasites, without eyes, without apparatus of motion, and without antennae. The lively, active form of youth, has become a shapeless, motionless lump. Only the most necessary organs of nutrition and propagation retain their activity; all the rest of the body has degenerated. Evidently these complete transformations are, to a large extent, the direct consequences of cumulative adaption, of the non-use and defective exercise of the organs, but a great portion of them must certainly be attributed also to correlative adaptation. (Compare Plate X. and XI.)

A seventh law of adaptation, the fourth in the group of direct adaptation, is *the law of divergent adaptation*. By this law we indicate the fact that parts originally formed alike have developed in different ways under the influence of external conditions. This law of adaptation is extremely important for the explanation of the phenomenon of division of labour, or polymorphism. We can see this very easily in our own selves; for instance, in the activity of our two hands. We usually accustom our right hand to quite different work from that which we give our left, and in consequence of the different occupation there arises a different formation of the two hands. The right hand, which we use much more than the left, shows a stronger development of the nerves, muscles, and bones. The same applies to the whole arm. In most human beings the bones and flesh of the right arm are, in consequence of their being more employed, stronger and heavier than those of the left arm. Now, as the special use of the right arm has been adopted and transmitted by inheritance for thousands of years among Europeans, the stronger shape and size of the right arm have already become hereditary. P. Harting, an excellent Dutch naturalist, has shown by measuring and weighing newly-born children, that even in them the right arm is more developed than the left.

According to the same law of divergent adaptation, both eyes also frequently develop differently. If, for example, a naturalist accustoms himself always to use one eye for the microscope (it is better to use the left), then that eye will acquire a power different from that of the other, and this division of labour is of great advantage. The one eye will become more short-sighted, and better suited for seeing things near at hand; the other eye becomes, on the contrary, more long-sighted, more acute for looking at an object in the distance. If, on the other hand, the naturalist alternately uses both eyes for the microscope, he will not acquire the short-sightedness of the one eye and the compensatory degree of long-sight in the other, which is attained by a wise distribution of these different functions of sight between the two eyes. Here then again the function, that is the activity, of originally equally-formed organs can become divergent by habit; the function reacts again upon the form of the organ, and thus we find, after a long duration of such an influence, a change in the more delicate parts and the relative growth of the divergent organs, which in the end becomes apparent even in their coarser outlines.

Divergent adaptation can very easily be perceived among plants, especially in creepers. Branches of one and the same creeping plant, which originally were formed alike, acquire a completely different form and extent, a completely different degree of curvature and diameter of spiral winding, according as they twine themselves round a thinner or a thicker bar. The divergent change of form of parts originally identical in form, which tending in different directions develop themselves under different external conditions, can be distinctly demonstrated in many other examples. As this divergent adaptation interacts with progressive inheritance, it becomes the cause of a division of labour among the different organs.

An eighth and last law of adaptation we may call *the law of unlimited or infinite adaptation.* By it we simply mean to express that we know of no limit to the variation of organic forms occasioned by the external conditions of existence. We can assert of no single part of an organism, that it is no longer variable, or that if it were subjected to new external conditions it would not be changed by them. It has never yet been proved by experience that there is a limit to variation. If, for example, an organ degenerates from non-use, this degeneration ends finally in a complete disappearance of the organ, as is the case with the eyes of many animals. On the other hand, we are able, by continual practice, habit, and the ever-increasing use of an organ, to bring it to a degree of perfection which we should at the beginning have considered to be impossible. If we compare the uncivilized savages with civilized nations, we find among the former a development of the organs of sense— sight, smell, and hearing—such as civilized nations can hardly conceive of. On the other hand, the brain, that is mental activity, among more civilized nations is developed to a degree of which the wild savages have no idea.

There appears indeed to be a limit given to the adaptability of every organism, by the "type" of its tribe or phylum; that is, by the essential fundamental qualities of this tribe, which have been inherited from a common ancestor, and transmitted by conservative inheritance to all its descendants. Thus, for example, no vertebrate animal can acquire the ventral nerve-chord of articulate animals, instead of the characteristic spinal marrow of the vertebrate animals. However, within this hereditary primary form, within this inalienable type, the degree of adaptability is unlimited. The elasticity and fluidity of the organic form manifests itself, within the type, freely in all directions, and to an unlimited extent. But there are some animals, as, for example, the parasitically degenerate crabs and worms, which seem to pass even the limit of type, and have forfeited all the essential characteristics of their tribe by an astonishing degree of degeneration. As to the adaptability of man, it is, as in all other animals, also unlimited, and since it is manifested in him above all other animals, in the modifications of the brain, there can be absolutely no limit to the knowledge which man in a further progress of mental cultivation may not be able to exceed. The human mind, according to the law of unlimited adaptation, enjoys an infinite perspective of becoming ever more and more perfect.

These remarks are sufficient to show the extent of the phenomena of Adaptation, and the great importance to be attached to them. The laws of Adaptation, or the facts of Variation caused by the influence of external conditions, are just as important as the laws of Inheritance. All phenomena of Adaptation, in the end, can be traced to conditions of nutrition of the organism, in the same way as the phenomena of Inheritance are referable to conditions of reproduction; but the latter, as well as the former, may further be traced to chemical and physical, that is to mechanical, causes. According to Darwin's Theory of Selection the new forms of organisms, the transformations which artificial selection produces in the state of cultivation, and which natural selection produces in the state of nature, arise solely by the interaction of such causes.

CHAPTER XI.

NATURAL SELECTION BY THE STRUGGLE FOR EXISTENCE. DIVISION OF LABOUR AND PROGRESS.

Interaction of the Two Organic Formative Causes, Inheritance and Adaptation.—Natural and Artificial Selection.—Struggle for Existence, or Competition for the Necessaries of Life.—Disproportion between the Number of Possible or Potential, and the Number of Real or Actual Individuals.—Complicated Correlations of all Neighbouring Organisms.— Mode of Action in Natural Selection.—Homochromic Selection as the Cause of Sympathetic Colourings.—Sexual Selection as the Cause of the Secondary Sexual Characters.— Law of Separation or Division of Labour (Polymorphism, Differentiation, Divergence of Characters).—Transition of Varieties into Species.—Idea of Species.—Hybridism.—Law of Progress or Perfectioning (Progressus, Teleosis).

In order to arrive at a right understanding of Darwinism, it is, above all, necessary that the two organic functions of *Inheritance and Adaptation,* which we spoke of in our last chapter, should be more closely examined. If we do not, on the one hand, examine the purely mechanical nature of these two physiological activities, and the various action of their different laws, and if, on the other hand, we do not consider how complicated the interaction of these different laws of Inheritance and Adaptation must be, we shall not be able to understand how these two functions, by themselves, have been able to produce all the variety of animal and vegetable forms, which, in fact, they have.

We have, at least, hitherto been unable to discover any other formative causes besides these two, and if we rightly understand the necessary and infinitely complicated interaction of Inheritance and Adaptation, we do not require to look for other unknown causes for the change of organic forms. These two fundamental causes are, as far as we can see, completely sufficient.

Even long before Darwin had published his Theory of Selection, some naturalists, and especially Goethe, had assumed the interaction of two distinct formative tendencies—a conservative or preserving, and a progressive or changing formative tendency—as the causes of the variety of organic forms. The former was called by Goethe the centripetal or specifying tendency, the latter the centrifugal tendency, or the tendency to metamorphosis (p. 89 [50]).

These two tendencies completely correspond with the two processes of Inheritance and Adaptation. *Inheritance* is the *centripetal* or *internal formative tendency* which strives to keep the organic form in its species, to form the descendants like the parents, and always to produce identical things from generation to generation. *Adaptation,* on the other hand, which counteracts inheritance, is the *centrifugal* or *external formative tendency,* which constantly strives to change the organic forms through the influence of the varying agencies of the outer world, to create new forms out of those existing, and entirely to destroy the constancy or permanency of species. Accordingly as Inheritance or Adaptation predominates in the struggle, the specific form either remains constant or changes into a new species. The degree of constancy of form in the different species of animals and plants, which obtains at any moment, is simply the necessary result of the momentary predominance which either of these two formative powers (or physiological activities) has acquired over the other.

If we now return to the consideration of the process of selection or choice, the outlines of which we have already examined, we shall be in a position to see clearly and distinctly that both artificial and natural selection rest solely upon the interaction of these two formative tendencies. If we carefully watch the proceedings of an artificial selector—a farmer or a gardener—we find that only these two constructive forces are used by him for the production of new forms. The whole art of artificial selection rests solely upon a thoughtful and wise application of the laws of Inheritance and Adaptation, and upon their being applied and regulated in an artistic and systematic manner. Here the will of man constitutes the selecting force.

The case of natural selection is quite similar, for it also employs merely these two organic constructive forces, these ingrained physiological properties of Adaptation and Heredity, in order to produce the different species. But the selecting principle or force, which in *artificial* selection is represented by the conscious *will of man* acting for a definite purpose, consists in *natural* selection of the unconscious *struggle for existence* acting without a definite plan. What we mean by "struggle for existence" has already been explained in the seventh chapter. It is the recognition of this exceedingly important identity which constitutes one of the greatest of Darwin's merits. But as this relation is very frequently imperfectly or falsely understood, it is necessary to examine it now more closely, and to illustrate by a few examples the operation of the struggle for life, and the operation of natural selection *by means of* the struggle for life (Gen. Morph. ii. 231).

When considering the struggle for life, we started from the fact that the number of germs which all animals and plants produce is infinitely greater than the number of individuals which actually come to life and remain alive for a longer or shorter time. Most organisms produce during life thousands or millions of germs, from each of which, under favourable circumstances, a new individual might arise. In most animals and plants these germs are eggs, that is cells, which for their development require sexual fructification. But among the Protista, the lowest organisms, which are neither animals nor plants, and which propagate themselves only in a non-sexual manner, the germ-cells, or spores, require no fructification. Now, in all cases the number of unsexual, as well as of sexual germs, is out of all proportion to the number of actually living individuals of every species.

Taken as a whole, the number of living animals and plants on our earth remains always about the same. The number of places in the economy of nature is limited, and in most parts of the earth's surface these places are always approximately occupied. Certainly there occur everywhere and in every year fluctuations in the absolute and in the relative number of individuals of all species. However, taken as a whole, these fluctuations are of little importance, and it is broadly the fact that the total number of all individuals remains, on an average, almost constant. There is a constant fluctuation, which depends on the fact that in one year or another one or other series of animals and plants predominates, and that every year the struggle for life somewhat alters their relations.

Every single species of animals and plants would have densely peopled the whole earth's surface in a short time, if it had not had to struggle against a number of enemies and hostile influences. Even Linnaeus calculated that if an annual plant only produced two seeds (and there is not one which produces so few), it would have yielded in twenty years a million of individuals. Darwin has calculated of elephants, which of all animals seem the slowest to increase, that in seven hundred and fifty years the descendants of a single pair would amount to nineteen millions of individuals; this is supposing that every elephant, during its period of fertility (from the 30th to the 90th year), produced only three pairs of young ones, and survived itself to its hundredth year. In like manner the increase of the number of human beings—if calculated on the average proportion of births to population, and no hindrances to the natural increase stood in the way—would be such as to double the total in twenty-five years. In every century the total number of men would have increased sixteen-fold; whereas we know that the total number of human beings increases but slowly, and that the increase of population is very different in different countries. While European tribes spread over the whole globe, other tribes or species of men every year draw nearer to their complete extinction. This is the case especially with the red-skins of America, and with the copper-coloured natives of Australia. Even if these races were to propagate more abundantly than the white Europeans, yet they would sooner or later succumb to the latter in the struggle for life. But of all human individuals, as of all other organisms, by far the majority perish at the earliest period of their lives. Of the immense quantity of germs which every species produce, only very few actually succeed in developing, and of these few it is again only a very small portion which attain to the age in which they can reproduce themselves (compare p. 161 [92]).

From the disproportion between the immense excess of organic germs and the small number of chosen individuals which are actually able to continue in existence beside one another, there follows of necessity that universal struggle for life, that constant fight for existence, that perpetual competition for the necessaries of life, of which I gave a sketch in my seventh chapter. It is this struggle for life which brings natural selection into play, which in its turn is made use of by the interaction of the phenomena of Inheritance and Adaptation as a sifting agency, and which thus causes a continual change in all organic forms. In this struggle for acquiring the necessary conditions of existence, those individuals will always overpower their rivals who possess any individual privilege, any advantageous quality, of which their fellow competitors are destitute. It is true we are able only in the fewest cases (in those animals and plants best known to us) to form an approximate conception of the infinitely complicated interaction of the numerous circumstances, all of which here come into combination. Only think how infinitely varied and complicated are the relations of every single human being to the rest of mankind, and in general, to the whole of the surrounding outer world. But similar relations prevail also among all animals and plants which live together in one place. All influence one another actively or passively. Every animal and every plant struggles directly with a number of enemies, beasts of prey, parasitic animals, etc. Plants standing together struggle with one another for the space of ground requisite for their roots, for the necessary amount of light, air, moisture, etc. In like-manner, animals living together struggle with one another for their food, dwelling-place, etc. In this most active and complicated struggle, any personal superiority, however small, any individual advantage, may possibly decide the issue in favour of the one possessing it. This privileged individual remains the victor in the struggle, and propagates itself, while its fellow-competitors perish before they succeed in propagating themselves. The personal advantage which gave it the victory is transmitted by inheritance to its descendants, and by a further development may become so strongly marked as to cause us to consider the later generations as a new species.

The infinitely complicated correlations which exist between the organisms of every district, and which must be looked upon as the real conditions of the struggle for life, are mostly unknown to us, and are very difficult to discover. We have hitherto been able to trace them only to a certain point in individual cases, as in the example given by Darwin of the relations between cats and red clover in England. The red clover (*Trifolium pratense*), which in England is among the best fodder for cattle, requires the visit of humming-bees in order to attain the formation of seeds. These insects, while sucking the honey from the bottom of the flower, bring the pollen in contact with the stigma, and thus cause the fructification of the flower, which never takes place without it. Darwin has shown by experiments, that red clover which is not visited by humming-bees does not yield a single seed. The number of bees is determined by the number of their enemies, the most destructive of which are the field-mice. The more the field-mice predominate, the less the clover is fructified. The number of field-mice, again, is dependent upon the number of their enemies, principally cats. Hence in the neighbourhood of villages and towns, where many cats are kept, there are plenty of bees. A great number of cats, therefore, is evidently of great advantage for the fructification of clover. This example may be followed still further, as has been done by Carl Vogt, if we consider that cattle which feed on red clover are one of the most important foundations of the wealth of England. Englishmen preserve their bodily and mental powers chiefly by making excellent meat—roast beef and beefsteak—their principal food. The English owe the superiority of their brains and minds over those of other nations in a great measure to their excellent meat. But this is clearly indirectly dependent upon the cats, which pursue the mice. We may, with Huxley, even trace the chain of causes to those old maids who cherish and keep cats, and, consequently, are of the greatest importance to the fructification of the clover and to the prosperity of England. From this example we can see that the further it is traced the wider is the circle of action and of correlation. We can with certainty maintain that there exist a great number of such correlations in every plant and in every animal, only we are not always able to point out and survey their concatenation as in the last instance.

Another remarkable example of important correlations is the following, given by Darwin. In Paraguay, there are no wild oxen and horses, as in the neighbouring parts of South America, both north and south of Paraguay. This surprising circumstance is explained simply by the fact that in that country a kind of small fly is very frequent, and is in the habit of laying its eggs in the navel of newly-born calves and foals. The newly-born animals die in consequence of this attack, and the small deadly fly is therefore the cause of oxen and horses never becoming wild in that district. Supposing that this fly were destroyed by some insect-eating bird, then these large mammals would grow wild in Paraguay, as well as in the neighbouring parts of South America; and as they would eat a quantity of certain species of plants, the whole flora, and, consequently again, the whole fauna of the country would become changed. It is hardly necessary to state, that at the same time the whole economy, and consequently the character, of the human population would alter.

Thus the prosperity, nay, even the existence of whole populations can be indirectly determined by a single small animal or vegetable form in itself extremely insignificant. There are small coral islands whose human inhaitants live almost entirely upon the fruit of a species of palm. The fructification of this palm is principally effected by insects, which carry the pollen from the male to the female palm trees. The existence of these useful insects is endangered by insect-eating birds, which in their turn are pursued by birds of prey. The birds of prey, however, often succumb to the attack of a small parasitical mite, which develops itself in millions in their feathers. This small, dangerous parasite, again, may be killed by parasitical moulds. Moulds, birds of prey, and insects would in this case favour the prosperity of the palm, and consequently of man; birds, mites, and insect-eating birds would, on the other hand, endanger it.

Interesting examples in relation to the change of correlations in the struggle for life are furnished also by those isolated oceanic islands, uninhabited by man, on which at different times goats and pigs have been placed by navigators. These animals become wild, and having no enemies, they increase in number so excessively, that the rest of the animal and vegetable population suffer in consequence, and the island finally may become almost a waste, because there is insufficient food for the large mammals which increase too numerously. In some cases on an island thus overrun with goats and pigs, other navigators have let loose a couple of dogs, who enjoyed this superabundance of food, and they again increased so numerously, and made such havoc among the herds, that after several years the dogs themselves lacked food, and they also almost died out. The equilibrium of species continually changes in this manner in nature's economy, accordingly as one or another species increases at the expense of the rest. In most cases the relations of different species of animals and plants to one another are much too complicated for us to be able to follow them, and I leave it to the reader to picture to himself what an infinitely complicated machinery is at work in every part of the world in consequence of this struggle. The impulses which started the struggle, and which altered and modified it in different places, are in the end seen to be the impulses of self-preservation—in fact, the instinct leading individuals to preserve themselves (the instinct of obtaining food), and the instinct leading them to preserve the species (instinct of propagation). It is these two fundamental instincts of organic self-preservation of which Schiller, the idealist (not Goethe, the realist!) says:

> "Meanwhile, until philosophy
> Sustains the structure of the world,
> Her workings will be carried on
> By hunger and by love."*

It is these two powerful fundamental instincts which, by their varying activity, produce such extraordinary differences in species through the struggle for life. They are the foundations of the phenomena of Inheritance and Adaptation. We have, in fact, traced all phenomena of Inheritance to

---

* Einstweilen bis den Bau der Welt
  Philosophie zusammenhält
  Erhält sich ihe Getriebe
  Durch Hunger und durch Liebe.

propagation, all phenomena of Adaptation to nutrition, as the two wider classes of material phenomena to which they belong.

The struggle for life in natural selection acts with as much selective power as does the will of man in artificial selection. The latter, however, acts according to a plan and consciously, the former without a plan and unconsciously. This important difference between artificial and natural selection deserves especial consideration. For we learn by it to understand how *arrangements serving a purpose can be produced by mechanical causes acting without an object, as well as by causes acting for an object.* The products of natural selection are arranged even more for a purpose than the artificial products of man, and yet they owe their existence not to a creative power acting for a definite purpose, but to a mechanical relation acting unconsciously and without a plan.

If we had not thoroughly considered the interaction of Inheritance and Adaptation under the influence of the struggle for life, we should not at first be inclined to expect such results from this natural process of selection as are, in fact, furnished by it. It may therefore be appropriate here to mention a few especially striking examples of the activity of natural selection.

Let us first take *Darwin's homochromic selection* of animals, or the so-called "sympathetic selection of colours," into consideration. Earlier naturalists have remarked that numerous animals are of nearly the same colour as their dwelling-place, or the surroundings in which they permanently live. Thus, for example, plant-lice and many other insects living on leaves are of a green colour. The inhabitants of the deserts, the jerboa, or leaping mice, foxes of the desert, gazelles, lions, etc., are mostly of a yellow or yellowish-brown colour, like the sand of the desert. The polar animals, which live on the ice and snow, are white or grey, like ice and snow. Many of these animals change their colour in summer and winter. In summer, when the snow partly vanishes, the fur of these polar creatures becomes brownish-grey or blackish, like the naked earth, while in winter it again becomes white. Butterflies and insects which hover round the gay and bright flowers are like them in colour. Now, Darwin explains this surprising circumstance quite simply by the fact that such colours as agree with the colour of the habitation are of the greatest use to the animals concerned. If these animals are animals of prey, they will be able to approach the object of their pursuit more safely and with less likelihood of observation, and, in like manner, those animals which are pursued will be able to escape more easily, if their colour is as little different as possible from that of their surroundings. If therefore originally an animal species varied so as to present cases of all colours, those individuals whose colour most resembled the surroundings must have been most favoured in the struggle for life. They remained more unobserved, maintained and propagated themselves, while those individuals or varieties differently coloured died out.

I have tried to explain, by the same sympathetic selection of colour, the wonderful fact that the majority of pelagic animals—that is, of those which live on the surface of the open sea—are bluish, or completely colourless and transparent, like glass and water itself. Such colourless, glassy animals are met with in the most different classes. To them belong, among fish, the Helmicthyidae, through whose crystalline bodies the words of a book can be read; among the molluscs, the finned snails (Heteropods) and sea-butterflies, or whales-food (Pteropods); among worms, the *Salpae*, *Alciope*, and *Sagitta*; further, a great number of pelagic crabs (Crustacea), and the greater part of the Medusae Umbrella-jellies, (Discomedusae); Comb-jellies, (Ctenophora). All of these pelagic animals, which float on the surface of the ocean, are transparent and colourless, like glass and like the water itself, while their nearest kin live at the bottom of the ocean, and are coloured and opaque like the inhabitants of the land. This remarkable fact, like the sympathetic colouring of the inhabitants of the earth, can be explained by natural selection. Among the ancestors of the pelagic glass-like animals which showed a different degree of colourlessness and transparency, those that were the most colourless and transparent must have been most favoured in the active struggle for life which takes place on the surface of the ocean. They were enabled to approach their prey the most easily unobserved, and were themselves least observed by their enemies. Hence they could preserve and propagate themselves more easily than their more coloured and opaque relatives; and finally, by accumulative adaptation and transmission by inheritance, through natural selection, in the course of many generations their bodies would attain that degree of crystal-like transparency and colourlessness which we at present admire in them. (Gen. Morph. ii. 242.)

No less interesting and instructive than homochromic selection is that species of natural selection which Darwin calls "*sexual selection*," which explains the origin of the so-called "secondary sexual characters." We have already mentioned these subordinate sexual characteristics, so instructive in many respects. They comprise those peculiarities of animals and plants which belong only to one of the two sexes, and which do not stand in any direct relation to the act propagation itself (compare above, p. 244 [139]). Such secondary sexual characters occur in great variety among animals. We all know how striking is the difference of the two sexes in size and colour in many birds and butterflies. The male sex is generally the larger and more beautiful. It often possesses special decorations or weapons; as for example, the spur and comb of the cock, the antlers of the stag and deer, etc. All these peculiarities of the two sexes have nothing directly to do with propagation itself, which is effected by the "primary sexual characters," or actual sexual organs.

Now, the origin of these remarkable "secondary sexual characters" is explained by Darwin simply by a choice or selection which takes place in the propagation of animals. In most animals the number of individuals of both sexes is unequal; either the number of the female or the number of the male individuals is greater, and, as a rule, when the season of propagation approaches, a struggle takes place between the rivals for the possession of the animals of the other sex. It is well known with what vigour and vehemence this struggle is fought out among the higher animals—among mammals and birds—especially among those of polygamous habits. Among gallinaceous birds, where for one cock there are several hens, a severe struggle takes place between the competing cocks for as large a harem as possible. The same is the case with many ruminating animals. Among stags and deer, for instance, at the period of rut, deadly struggles take place between the males for the possession of the females. The secondary sexual character which here distinguishes the males—the antlers of stags and deer—not possessed by the female, is, according to Darwin, the consequence of that struggle. Here the motive and cause determining the struggle is not, as in the case of the struggle for individual existence, self-preservation, but the preservation of the species—propagation. There are numerous passive weapons of defence, as well as active weapons for attack. The lion's mane, not possessed by the female, is evidently such a weapon of defence; it is an excellent means of protection against the bites which the male lions try to inflict on each other's necks when fighting for the females; consequently those males with the strongest manes have the greatest advantage in the sexual struggle. The dewlap of the ox and the comb of the cock are similar defensive weapons. Active weapons of attack, on the other hand, are the antlers of the stag, the tusks of the boar, the spur of the cock, and the hugely developed pair of jaws in the male stag-beetle; all are instruments employed by the males in the struggle for the females, for annihilating or chasing away their rivals.

In the cases just mentioned, it is the bodily "struggle to the death" which determines the origin of the secondary sexual characters. But, besides these mortal struggles, there are other important competitions in sexual selection, which no less influence the structure of the rivals. These consist principally in the fact that the courting sex tries to please the other by external finery, by beauty of form, or by a melodious voice. Darwin thinks that the beautiful voices of singing birds have principally originated in this way. Many male birds carry on a regular musical contest when they contend for the possession of the females. It is known of several singing birds, that in the breeding season the males assemble in numbers round the females, and let their songs resound before them, and that then the females choose the singers who best please them for their mates. Among other songsters, individual males pour out their songs in the loneliness of the forest in order to attract the females, and the latter follow the most attractive calls. A similar musical contest, though certainly less melodious, takes place among crickets and grasshoppers. The male cricket has on its belly two instruments like drums, and produces with these the sharp chirping notes which the ancient Greeks curiously enough thought beautiful music. Male grasshoppers, partly by using their hind-legs like the bow of a violin against their wing coverings, and partly by rubbing their wing coverings together, bring out tones which are, indeed, not melodious to us, but which please the female grasshoppers so much that they choose the male who fiddles the best.

Among other insects and birds it is not song or, in fact, any musical accomplishment, but finery or beauty of the one sex which attracts the other. Thus we find that, among most gallinaceous birds, the cocks are distinguished by combs on their heads, or by a beautiful tail, which they can spread out like a fan; as for example, in the case of the peacock and turkey-cock. The magnificent tail of the bird of paradise is also an exclusive ornament of the male sex. In like manner, among very many other birds and very many insects, principally among butterflies, the males are distinguished from the females by special colours or other decorations. These are evidently the results of sexual selection. As the females do not possess these attractions and decorations, we must come to the conclusion that they have been acquired by degrees by the males in the competition for the females, which takes its origin in the selective discrimination of the females.

We may easily picture to ourselves, in detail, the application of this interesting conclusion to the human community. Here, also, the same causes have evidently influenced the development of the secondary sexual characters. The characteristics distinguishing the man, as well as those distinguishing the woman, owe their origin, certainly for the most part, to the sexual selection of the other sex. In antiquity and in the Middle Ages, especially in the romantic age of chivalry, it was the bodily struggles to the death—the tournaments and duels—which determined the choice of the bride; the strongest carried home the bride. In more recent times, however, in our so-called "polished" or "highly civilized" society, competing rivals prefer to contend indirectly by means of musical accomplishments, instrumental performances and song, by bodily charms, natural beauty, or artificial decoration. But by far the most important of these different forms of sexual selection in man is that form which is the most exalted, namely, *psychical selection*, in which the mental excellencies of the one sex influence and determine the choice of the other. The most highly intellectually developed types of men have, throughout generations, when choosing a partner in life, been guided by her excellencies of soul, and have thus transmitted these qualities to their posterity, and they have in this way, more than by any other thing, helped to create the deep chasm which at present separates civilized men from the rudest savages, and from our common animal ancestors. In fact, both the part played by the prevalence of a higher standard of sexual selection, and the part played by the due division of labour between the two sexes, is exceedingly important, and I believe that here we must seek for the most powerful causes which have determined the origin and the historical development of the races of man. (Gen. Morph. ii. 247.) As Darwin, in his exceedingly interesting work, published in 1871, on "The Origin of Man and

Sexual Selection,"[48] has discussed this subject in the most masterly manner, and has illustrated it by most remarkable examples, I refer for further detail to that work.

But now let us look again at two extremely important organic laws which can be explained by the theory of selection, as necessary consequences of natural selection in the struggle for existence. I mean the law of *division of labour*, or *differentiation*, and the law of *progress*, or *perfecting*. When the phenomena due to these two laws first became known, through observation of the historical development, the individual development, and the comparative anatomy of animals and plants, naturalists were inclined to trace them to a direct creative influence. It was supposed to be part of the plan of the Creator, acting for a definite purpose, in the course of time to develop the forms of animals and plants more and more variously, and to bring them more and more to a state of perfection. We shall evidently make a great advance in the knowledge of nature if we reject this teleological and anthropomorphic conception, and if we can prove the two laws of Division of Labour and Perfecting to be the necessary consequences of natural selection in the struggle for life.

The first great law which follows directly and of necessity from natural selection, is that of *separation*, or *differentiation*, which is frequently called *division of labour*, or *polymorphism*, and which Darwin speaks of as *divergence of character*. (Gen. Morph. ii. 249.) We understand by it the general tendency of all organic individuals to develop themselves more and more diversely, and to deviate from the common primary type. The cause of this general inclination towards differentiation and the formation of heterogeneous forms from homogeneous beginnings is, according to Darwin, simply to be traced to the circumstance that the struggle for life between every two organisms rages all the more fiercely the nearer the relation in which they stand to one another, or the more nearly alike they are. This is an exceedingly important, and in reality an exceedingly simple relation, but it is usually not duly considered.

It must be obvious to every one, that in a field of a certain size, beside the corn-plants which have been sown, a great number of weeds can exist, and, moreover, in places which could not have been occupied by corn-plants. The more dry and sterile places of the ground, in which no corn-plant would thrive, may still furnish sustenance to weeds of different kinds; and such species and individuals of weeds will more readily be able to exist in such conditions, in proportion as they are suited to adapt themselves to the different parts of the ground. It is the same with animals. It is evident that a much greater number of animal individuals can live together in one and the same limited district, if they are of various and different natures, than if they are all alike. There are trees (for example, the oak) on which a couple of hundred of different species of insects live together. Some feed on the fruits of the tree, others on the leaves, others again on the bark, the root, etc. It would be quite impossible for an equal number of individuals to live on this tree if all were of one species; if, for example, all fed on the bark, or only upon the leaves. Exactly the same is the case in human society. In one and the same small town, only a certain number of workmen can exist, even when they follow different occupations. The division of labour, which is of the greatest use to the whole community, as well as to the individual workman, is a direct consequence of the struggle for life, of natural selection; for this struggle can be sustained more easily the more the activities, and hence, also, the forms of the different individuals deviate from one another. The different function naturally produces its reaction in changing the form, and the physiological division of labour necessarily determines the morphological differentiation, that is, the "divergence of character."[37]

Now, I beg the reader again to remember that all species of animals and plants are variable, and possess the capability of adapting themselves to different places or to local relations. The varieties or races of each species, according to the laws of adaptation, deviate all the more from the original primary species, the greater the difference of the new conditions to which they adapt themselves. If we imagine these varieties—which have proceeded from a common primary form—to be disposed in the shape of a branching, radiating bunch, then those varieties will be best able to exist side by side and propagate which are most distant from one another, which stand at the ends of the series, or at the opposite sides of the bunch. Those forms, on the other hand, occupying a middle position—presenting a state of transition—have the most difficult position in the struggle for life. The necessaries of life differ most in the two extremes, in the varieties most distant from one another, and consequently these will get into the least serious conflict with one another in the general struggle for life. But the intermediate forms, which have deviated less from the original primary form, require nearly the same necessaries of life as the original form, and therefore, in competing for them, they will have to struggle most with, and be most seriously threatened by, its members. Consequently, when numerous varieties of a species live side by side on the same spot of the earth, the extremes, or those forms deviating most from one another, can much more easily continue to exist beside one another than the intermediate forms which have to struggle with each of the different extremes. The intermediate forms will not be able to resist, for any length of time, the hostile influences which the extreme forms victoriously overcome. These alone maintain and propagate themselves, and at length cease to be any longer connected with the original primary species through intermediate forms of transition. Thus arise "good species" out of varieties. Thus, then, the struggle for life necessarily favours the general divergence of organic forms, that is, the constant tendency of organisms to form new species. This fact does not rest upon any mystic quality, or upon an unknown formative tendency, but upon the interaction of Inheritance and Adaptation in the struggle for life. As the intermediate forms, that is, the individuals in a state of transition, of the varieties of every species die out and become extinct, the process of divergence constantly goes further, and from the extremes forms develop which we distinguish as new species. Although all naturalists have been obliged to acknowledge the variability and mutability of all species of animals and plants, yet most of them have hitherto denied that the modification or transformation of the organic form surpasses the original limit of the characters of the species. Our opponents cling to the proposition—"However far a species may exhibit deviations from its usual form in a collection of varieties, yet the varieties of it are never so distinct from one another as two really good species." This assertion, which Darwin's opponents usually place at the head of their arguments, is utterly untenable and unfounded. This will become quite clear as soon as we critically compare the various attempts to define the idea of species. No naturalist can answer the question as to what is in reality a "genuine or good species" ("bona species"); yet every systematic naturalist uses this expression every day, and whole libraries have been written on the question as to whether this or that observed form is a species or a variety, whether it is a really good or a bad species. The most general answer to this question used to be the following: "To one species belong all those individuals which agree in all essential characteristics. Essential characteristics of species are those which remain permanent or constant, and never become modified or vary." But as soon as a case occurred in which the characteristic—which had hitherto been considered essential—did become modified, then it was said, "This characteristic is not essential to the species, for essential characteristics never vary." Those who argued thus evidently moved in a circle, and the naivete with which this circular method of defining species is laid down in thousands of books as an unassailable truth, and is still constantly repeated, is truly astonishing.

All other attempts which have been made to arrive at a definite and logical determination of the idea of organic "species" have, like the last, been utterly futile, and led to no results. Considering the nature of the case, it cannot be otherwise. The idea of species is just as truly a relative one and not absolute, as is the idea of variety, genus, family, order, class, etc. I have proved this in detail in the criticism of the idea of species in my "General Morphology" (Gen. Morph. ii. 323–364). I will waste no more time on this unsatisfactory discussion, and now only add a few words about the *relation of species to hybridism*. Formerly it was regarded as a dogma, that two good species could never produce hybrids which could reproduce themselves as such. Those who thus dogmatized almost always appealed to the hybrids of a horse and donkey, the mule and the hinny, which, truly enough, are seldom able to reproduce themselves. But the truth is that such unfruitful hybrids are rare examples, and in the majority of cases hybrids of two totally different species are fruitful and able to reproduce themselves. They can almost always fruitfully mix with one or other of the parent species, and sometimes also among themselves; and in this way completely new forms can originate according to the laws of "mixed transmission by inheritance."

Thus, in fact, *hybridism is a source of the origin of new species*, distinct from the source we have hitherto considered—natural selection. I have already spoken occasionally of these *hybrid species* (species hybridae), especially of the hare-rabbit (*Lepus darwinii*), which has arisen from the crossing of a male hare and a female rabbit; the goat-sheep (*Capra ovina*), which has arisen from the pairing of a he-goat and ewe; also the different species of thistles (*Cirsium*), brambles (*Rubus*), etc. It is possible that many wild species have originated in this way, as even Linnaeus assumed. At all events, these hybrid species, which can maintain and propagate themselves as well as pure species, prove that hybridism cannot serve in any way to give an absolute definition to the idea of species.

I have already mentioned (p. 47 [27]) that the many vain attempts to define the idea of species theoretically have nothing whatever to do with the practical distinction of species. The extensive practical application of the idea of species, as it is carried out in systematic zoology and botany, is very instructive as furnishing an example of human folly. Hitherto, by far the majority of zoologists and botanists, in distinguishing and describing the different forms of animals and plants, have endeavoured, above all things, to distinguish accurately kindred forms as so many "good species." However, it has been found scarcely possible, in any group, to make an accurate and consistent distinction of such "genuine or good species." There are no two zoologists, no two botanists, who agree in all cases as to which of the nearly related forms of a genus are good species, and which are not. All authors have different views about them. In the genus *Hieracium*, for example, one of the commonest genera of European plants, no less than 300 species have been distinguished in Germany alone. The botanist Fries, however, only admits 106, Koch only 52, as "good species," and others accept scarcely 20. The differences in the species of brambles (*Rubus*) are equally great. Where one botanist makes more than a hundred species, a second admits only about one half of that number, a third only five or six, or even fewer species. The birds of Germany have long been very accurately known. Bechstein, in his careful "Natural History of German Birds," has distinguished 367 species, L. Reichenbach 379, Meyer and Wolff 406, and Brehm, a clergyman learned in ornithology, distinguishes even more than 900 different species.

Thus we see that here, and, in fact, in every other domain of systematic zoology and botany, the most arbitrary proceedings prevail, and, from the nature of the case, must prevail. For it is quite impossible accurately to distinguish varieties and races from so-called "good species." *Varieties are commencing species*. The variability or adaptability of species, under the influence of the struggle for life, necessitates the continual and progressive separation or differentiation of varieties, and the perpetual delimitation of new forms. Whenever these are maintained throughout a number of generations by inheritance, whilst the intermediate forms die out, they form independent "new species." The origin of new species by division of labour, or separation, divergence, or differentiation of varieties, is therefore a *necessary consequence of natural selection.*[37]

The same kind of interest attaches to a second great law which we deduce from natural selection, and which is, indeed, closely connected with the law of Divergence, but in no way identical with it; namely, the law of *Progress* (progressus), or *Perfecting* (teleosis). (Gen. Morph. ii. 257.) This great and important law, like the law of differentiation, had long been empirically established by palaeontological experience, before Darwin's Theory of Selection gave us the key to the explanation of its cause. The most distinguished palaeontologists have pointed out the law of progress as the most general result of their investigations of fossil organisms. This has been specially done by Bronn, whose investigations on the laws of construction[18] and the laws of the development[19] of organisms, although little heeded, are excellent, and deserve most careful consideration. The general results of the law of differentiation and the law of progress, at which Bronn arrived by a purely mechanical hypothesis, and by exceedingly accurate, laborious, and careful investigations, are brilliant confirmations of the truth of these two great laws which we deduce as necessary inferences from the theory of selection.

The law of progress or of perfecting establishes the exceedingly important fact, on the ground of palaeontological experience, that in successive periods of this earth's history, a continual increase in the perfection of organic formations has taken place. Since that inconceivably remote period in which life on our planet began with the spontaneous generation of Monera, organisms of all groups, both collectively as well as individually, have continually become more perfectly and highly developed. The steadily increasing variety of living forms has always been accompanied by progress in organization. The lower the strata of the earth in which the remains of extinct animals and plants lie buried, that is, the older the strata are, the more simple and imperfect are the forms which they contain. This applies to organisms collectively, as well as to every single large or small group of them, setting aside, of course, those exceptions which are due to the process of degeneration, which we shall discuss hereafter.

As a confirmation of this law I shall mention only the most important of all animal groups, the tribe of vertebrate animals. The oldest fossil remains of vertebrate animals known to us belong to the lowest class, that of Fishes. Upon these there followed later more perfect Amphibious animals, then Reptiles, and lastly, at a much later period, the most highly organized classes of vertebrate animals, Birds and Mammals. Of the latter only the lowest and most imperfect forms, without placenta, appeared at first, such as are the pouched animals (Marsupials), and afterwards, at a much later period, the more perfect mammals, with placenta. Of these, also, at first only the lower kinds appeared, the higher forms later; and not until the late tertiary period did man gradually develop out of these last.

If we follow the historical development of the vegetable kingdom we shall find the same law operative there. Of plants there existed at first only the lowest and most imperfect classes, the Algae or tangles. Later there followed the group of Ferns or Filicinae (ferns, pole-reeds, scale-plants, etc.). But as yet there existed no flowering plants, or Phanerogama. These originated later with the Gymnosperms (firs and cycads), whose whole structure stands far below that of the other flowering plants (Angiosperms), and forms the transition from the group of fern-like plants to the Angiosperms. These latter developed at a still later date, and among them there were at first only flowering plants without corolla (Monocotyledons and Monochlamyds); only later were there flowering plants with a corolla (Dichlamyds). Finally, again, among these the lower polypetalous plants preceded the higher gamopetalous plants. The whole series thus constitutes an irrefutable proof of the great law of progressive development.

Now, if we ask what is the cause of this fact, we again, just as in the case of differentiation, come back to natural selection in the struggle for life. If once more we consider the whole process of natural selection, how it operates through the complicated interaction of the different laws of Inheritance and Adaptation, we shall recognize not only divergence of character, but also the perfecting of structure to be the direct and necessary result of it. We can trace the same thing in the history of the human race. Here, too, it is natural and necessary that the progressive division of labour constantly furthers mankind, and urges every individual branch of human activity into new discoveries and improvements. This progress itself universally depends on differentiation, and is consequently, like it, a direct result of natural selection in the struggle for life.

CHAPTER XII.

LAWS OF DEVELOPMENT OF ORGANIC TRIBES AND OF INDIVIDUALS. PHYLOGENY AND ONTOGENY.

Laws of the Development of Mankind: Differentiation and Perfecting.—Mechanical Cause of these two Fundamental Laws.—Progress without Differentiation, and Differentiation without Progress.—Origin of Rudimentary Organs by Non-use and Discontinuance of Habit.—Ontogenesis, or Individual Development of Organisms.—Its General Importance.—Ontogeny, or the Individual History of Development of Vertebrate Animals, including Man.—The Fructification of the Egg.—Formation of the three Germ Layers.—History of the Development of the Central Nervous System, of the Extremities, of the Branchial Arches, and of the Tail of Vertebrate Animals.—Causal Connection and Parallelism of Ontogenesis and Phylogenesis, that is of the Development of Individuals and Tribes.—Causal Connection of the Parallelism of Phylogenesis and of Systematic Development.—Parallelism of the three Organic Series of Development.

IF man wishes to understand his position in nature, and to comprehend as natural facts his relations to the phenomena of the world cognisable by him, it is absolutely necessary that he should compare human with extra-human phenomena, and, above all, with animal phenomena. We have already seen that the exceedingly important physiological laws of Inheritance and Adaptation apply to the human organism in the same manner as to the animal and vegetable kingdoms, and in both cases interact with one another. Consequently, natural selection in the struggle for life acts so as to transform human society, just as it modifies animals and plants, and in both cases constantly produces new forms. The comparison of the phenomena of human and animal transformation is especially interesting in connection with the laws of divergence and progress, the two fundamental laws which, at the end of the last chapter, we proved to be direct and necessary consequences of natural selection in the struggle for life.

A comparative survey of the history of nations, or what is called "universal history," will readily yield to us, as the first and most general result, evidence of a continually *increasing variety* of human activities, both in the life of individuals and in that of families and states. This differentiation or separation, this constantly increasing divergence of human character and the form of human life, is caused by the ever advancing and more complete division of labour among individuals. While the most ancient and lowest stages of human civilization show us throughout the same rude and simple conditions, we see in every succeeding period of history, among different nations, a greater variety of customs, practices, and institutions. The increasing division of labour necessitates an increasing variety of forms corresponding to it. This is expressed even in the formation of the human face. Among the lowest tribes of nations, most of the individuals resemble one another so much that European travellers often cannot distinguish them at all. With increasing civilization the physiognomy of individuals becomes differentiated, and finally, among the most highly civilized nations, the English and Germans, the divergence in the characters of the face is so great that we very rarely mistake one face for another.

The second great fundamental law which is obvious in the history of nations is the great law of progress or perfecting. Taken as a whole, the history of man is the history of his *progressive development*. It is true that everywhere and at all times we may notice individual retrogressions, or observe that crooked roads towards progress have been taken, which lead only towards one-sided and external perfecting, and thus deviate more and more from the higher goal of internal and enduring perfecting. However, on the whole, the movement of development of all mankind is and remains a progressive one, inasmuch as man continually removes himself further from his ape-like ancestors, and continually approaches nearer to his own ideal.

Now, if we wish to know what causes laws of development in man, namely, progress, we must compare them with development in animals, and on a close examination we shall inevitably come to the conclusion that the phenomena, as well as their causes, are exactly the same in the two cases. The course of development in man, just as in that of animals, being directed by the two fundamental laws of differentiation and perfecting, is determined solely by purely mechanical causes, and is solely the necessary consequence of natural selection in the struggle for life.

Perhaps in the preceding discussion the question has presented itself to some—"Are not these two laws identical? Is not progress in all cases necessarily connected with divergence?" This question has often been answered in the affirmative, and Carl Ernst Bär, for example, one of the greatest investigators in the domain of the history of development, has set forth the following proposition as one of the principal laws in the ontogenesis of the animal body:—"The degree of development (or actually determine these two great the law of divergence and the law of the corresponding laws of perfecting) depends on the stage of separation (or differentiation) of the parts."[20] Correct as this proposition may be on the whole, yet it is not universally true. In many individual cases it can be proved that divergence and progress by no means always coincide. *Every progress is not a differentiation, and every differentiation is not a progress.*

Naturalists, guided by purely anatomical considerations, had already set forth the law relating to progress in organization, that the perfecting of an organism certainly depends, for the most part, upon the division of labour among the individual organs and parts of the body, but that there are also other organic transformations which determine a progress in organization. One, in particular, which has been generally recognized, is the *numerical diminution of identical parts*. If, for example, we compare the lower articulated animals of the crustacean group, which possess numerous pairs of legs, with spiders which never have more than four pairs of legs, and with insects which always possess only three pairs of legs, we find this law, for which a great number of examples could be adduced, confirmed. The numerical diminution of pairs of legs is a progress in the organization of articulated animals. In like manner the numerical diminution of corresponding vertebral joints in the trunk of vertebrate animals is a progress in their organization. Fishes and amphibious animals with a very large number of identical vertebral joints are, for this very reason, less perfect and lower than birds and mammals, in which the vertebral joints, as a whole, are not only very much more differentiated, but in which the number of corresponding vertebrae is also much smaller. Further, according to the same law of numerical diminution, flowers with numerous stamens are more imperfect than the flowers of kindred plants with a smaller number of stamens, etc. If therefore originally a great number of homogeneous parts exist in an organic body, and if, in the course of very many generations, this number be gradually decreased, this transformation will be an example of perfecting.

Another law of progress, which is quite independent of differentiation, nay, even appears to a certain extent opposed to it, is the law of *centralization*. In general the whole organism is the more perfect the more it is organized as a unit, the more the parts are subordinate to the whole, and the more the functions and their organs are centralized. Thus, for example, the system of blood-vessels is most perfect where a centralized heart exists. In like manner, the dense mass of marrow which forms the spinal cord of vertebrate animals, and the ventral cord of the higher articulated animals, is more perfect than the decentralized chain of ganglia of the lower articulated animals, and the scattered system of ganglia in the molluscs. Considering the difficulty of explaining these complicated laws of progress in detail, I cannot here enter upon a closer discussion of them, and must refer to Bronn's excellent "Morphologischen Studien," and to my "General Morphology" (Gen. Morph. i. 370, 550; ii. 257–266).

Just as we have become acquainted with phenomena of progress, quite independent of divergence, so we shall, on the other hand, very often meet with divergencies which are not perfecting, but which are rather the contrary, that is retrogressions or degenerations. It is easy to see that the changes which every species of animal and plant experiences cannot always be improvements. But rather many phenomena of differentiation, which are of direct advantage to the organism itself, are yet, in a wider sense, detrimental, inasmuch as they lessen its general capabilities. Frequently a relapse to simpler conditions of life takes place, and by adaptation to them a divergence in a retrograde direction. If, for instance, organisms which have hitherto lived independently accustom themselves to a parasitical life, they thereby degenerate or retrograde. Such animals, which hitherto had possessed a well-developed nervous system and quick organs of sense, as well as the power of moving freely, lose these when they accustom themselves to a parasitical mode of life; they consequently retrograde more or less. There the differentiation viewed by itself is a degeneration, although it is advantageous to the parasitical organism. In the struggle for life such an animal, which has accustomed itself to live at the expense of others, by retaining its eyes and apparatus of motion, which are of no more use to it, would only expend so much material uselessly; and when it loses these organs, then a great quantity of nourishment which was employed for the maintenance of these parts, benefits other parts. In the struggle for life between the different parasites, therefore, those which make least pretensions will have advantage over the others, and this favours their degeneration.

Just as this is found to be the case with the whole organism, so it is also with the parts of the body of an individual organism. A differentiation of parts, which leads to a partial degeneration, and finally even to the loss of individual organs, is, when looked at by itself, a degeneration, but yet may be advantageous to the organism in the struggle for life. It is easier to fight when useless baggage is thrown aside. Hence we meet everywhere, in the more highly-developed animal and vegetable bodies, processes of divergence, the essence of which is that they cause the degeneration, and finally the loss, of particular parts. And at this point the most important and instructive of all the series of phenomena bearing upon the history of organisms presents itself to us, namely, that of *rudimentary or degenerate organs*.

It will be remembered that even in my first chapter I considered this exceedingly remarkable series of phenomena, from a theoretical point of view, as one of the most important and most striking proofs of the truth of the doctrine of descent. We designated as rudimentary organs those parts of the body which are arranged for a definite purpose and yet are without function. Let me remind the reader of the eyes of those animals which live in the dark in caves and underground, and which consequently never can use them. In these animals we find real eyes hidden under the skin, frequently developed exactly as are the eyes of animals which really see; and yet these eyes never perform any function, indeed cannot, simply for the reason that they are covered by an opaque membrane, and consequently no ray of light falls upon them (compare above, p. 13 [8]). In the ancestors of these animals, which lived in open daylight, the eyes were well developed, covered by a transparent horny capsule (cornea), and actually served the purpose of seeing. But as the animals gradually accustomed themselves to an underground mode of life, and withdrew from the daylight and no longer used their eyes, these became degenerated.

Very clear examples of rudimentary organs, moreover, are the wings of animals which cannot fly; for example, the wings of the running birds, like the ostrich, emu, cassowary, etc., the legs of which have become exceedingly developed. These birds having lost the habit of flying, have consequently lost the use of their wings; however, the wings are still there, although in a crippled form. We very frequently find such crippled wings in the class of insects, most members of which can fly.

From reasons derived from comparative anatomy and other circumstances, we can with certainty draw the inference that all insects now living (all dragon-flies, grasshoppers, beetles, bees, bugs, flies, butterflies, etc.) have originated from a single common parental form, from a primary insect which possessed two well-developed pairs of wings, and three pairs of legs. Yet there are very many insects in which either one or both pairs of wings have become more or less degenerated, and many in which they have even completely disappeared. For example, in the whole order of flies, or Diptera, the hinder pair of wings—in the bee-parasites, or Strepsiptera, on the other hand, the fore pair of wings—have become degenerated or entirely disappeared. Moreover, in every order of insects we find individual genera, or species, in which the wings have more or less degenerated or disappeared. The latter is the case especially in parasites. The females have frequently no wings, whereas the males have; for instance, in the case of glow-worms (*Lampyris*), Strepsiptera, etc. This partial or complete degeneration of the wings of insects has evidently arisen from natural selection in the struggle for life. For we find insects without wings living under circumstances where flying would be useless, or even decidedly injurious to them. If, for example, insects living on islands fly about much, it may easily happen that when flying they are blown into the sea by the wind, and if (as is always the case) the power of flying is differently developed in different individuals, then those which fly badly have an advantage over those which fly well; they are less easily blown into the sea, and remain longer in life than the individuals of the same species which fly well. In the course of many generations, by the action of natural selection, this circumstance must necessarily lead to a complete suppression of the wings. If this conclusion had been arrived at on purely theoretical grounds, we might be pleased to find its truth established by facts. For upon isolated islands the proportion of wingless insects to those possessing wings is surprisingly large, much larger than among the insects inhabiting continents. Thus, for example, according to Wollaston, of the 550 species of beetles which inhabit the island of Madeira, 220 are wingless, or possess such imperfect wings that they can no longer fly; and of the 29 genera which belong to that island exclusively, no less than 23 contain such species only. It is evident that this remarkable circumstance does not need to be explained by the special wisdom of the Creator, but is sufficiently accounted for by natural selection, because in this case the hereditary disuse of the wings, the discontinuance of flying in the presence of dangerous winds, has been very advantageous in the struggle for life. In other wingless insects the want of wings has been advantageous for other reasons. Viewed by itself, the loss of wings is a degeneration, but in these special conditions of life it is advantageous to the organism in the struggle for life.

Among other rudimentary organs I may here, by way of example, further mention the lungs of serpents and serpent-like lizards. All vertebrate animals possessing lungs, such as amphibious animals, reptiles, birds, and mammals, have a pair of lungs, a right and a left one. But in cases where the body is exceedingly thin and elongated, as in serpents and serpent-like lizards, there is no room for the one lung by the side of the other, and it is an evident advantage to the mechanism of respiration if only one lung is developed. A single large lung here accomplishes more than two small ones side by side would do; and consequently, in these animals, we invariably find only the right or only the left lung fully developed. The other is completely aborted, although existing as a useless rudiment. In like manner, in all birds the right ovary is aborted and without function; only the left one is developed, and yields all the eggs.

I mentioned in the first chapter that man also possesses such useless and superfluous rudimentary organs, and I specified as such the muscles which move the ears. Another of them is the rudiment of the tail which man possesses in his 3–5 tail vertebrae, and which, in the human embryo, stands out prominently during the first two months of its development (compare Plates II. and III.). It afterwards becomes completely hidden. The rudimentary little tail of man is an irrefutable proof of the fact that he is descended from tailed ancestors. In woman the tail is generally by one vertebra longer than in man. There still exist rudimentary muscles in the human tail which formerly moved it.

Another case of human rudimentary organs, only belonging to the male, and which obtains in like manner in all male mammals, is furnished by the mammary glands on the breast, which, as a rule, are active only in the female sex. However, cases of different mammals are known, especially of men, sheep, and goats, in which the mammary glands were fully developed in the male sex, and yielded milk as food for their offspring. I have already mentioned before (p. 12 [7]) that the rudimentary auricular muscles in man can still be employed to move their ears, by some persons who have perseveringly practised them. In fact, rudimentary organs are frequently very differently developed in different individuals of the same species; in some they are tolerably large, in others very small. This circumstance is very important for their explanation, as is also the other circumstance that generally in embryos, or in a very early period of life, they are much larger and stronger in proportion to the rest of the body than they are in fully developed and fully grown organisms. This can, in particular, be easily pointed out in the rudimentary sexual organs of plants (stamens and pistil), which I have already mentioned. They are proportionately much larger in the young flower-bud than in the mature flower.

I have remarked (p. 15 [9]) that rudimentary or suppressed organs were the strongest supports of the monistic or mechanical conception of the universe. If its opponents, the dualists and teleologists, understood the immense significance of rudimentary organs, it would put them into a state of despair. Their ludicrous attempts to explain that rudimentary organs were given to organisms by the Creator "for the sake of symmetry," or "as a formal provision," or "in consideration of his general plan of creation," sufficiently prove the utter impotence of their perverse conception of the universe. I must here repeat that, even if we knew absolutely nothing of the other phenomena of development, we should be obliged to believe in the truth of the Theory of Descent, solely on the ground of the existence of rudimentary organs. Not one of its opponents has been able to throw even a feeble glimmer of an acceptable explanation upon these exceedingly remarkable and important phenomena. There is scarcely any highly developed animal or vegetable form which has not some rudimentary organs, and in most cases it can be shown that they are the products of natural selection, and that they have become suppressed by disuse. It is the reverse of the process of formation in which new organs arise from adaptation to certain conditions of life, and by the use of parts as yet incompletely developed. It is true our opponents usually maintain that the origin of altogether new parts is completely inexplicable by the Theory of Descent. However, I distinctly assert that to those who possess a knowledge of comparative anatomy and physiology this matter does not present the slightest difficulty. Every one who is familiar with comparative anatomy and the history of development will find as little difficulty about the origin of completely new organs as about the utter disappearance of rudimentary organs. The disappearance of the latter, viewed by itself, is the converse of the origin of the former. Both processes are particular phenomena of differentiation, which, like all others, can be explained quite simply and mechanically by the action of natural selection in the struggle for life.

The infinitely important study of rudimentary organs and their origin, the comparison of their palaeontological and embryological development, now naturally leads us to the consideration of one of the most important and instructive of all biological phenomena, namely, the parallelism which the phenomena of progress and divergence present to us in three different series. When, in the last chapter, we spoke of perfecting and division of labour, we understood by those words progress and separation, and those changes effected by them, which in the long and slow course of the earth's history have led to a continual variation of the flora and fauna, to the origin of new and to the disappearance of ancient species of animals and plants. Now, if we follow the origin, the development, and the life of every single organic individual, we meet with exactly the same phenomena of progress and differentiation. The individual development, or the *ontogenesis* of every single organism, from the egg to the complete form is nothing but a growth attended by a series of diverging and progressive changes. This applies equally to animals, plants, and protista. If, for example, we consider the ontogeny of any mammal, of man, of an ape, or of a pouched animal, or if we follow the individual development of any other vertebrate animal of another class, we everywhere find essentially the same phenomena. Every one of these animals develops itself originally out of a single cell, the egg. This cell increases by self-division, and forms a number of cells, and by the growth of this accumulation of cells, by the divergent development of originally identical cells, by the division of labour among them, and by their perfecting, there arises the perfect organism, the complicated composition of which excites our admiration.

It seems to me here indispensable to draw attention more closely to those infinitely important and interesting processes which accompany *ontogenesis, or the individual development of organisms*, and especially to that of vertebrate animals, man included. I wish especially to recommend these exceedingly remarkable and instructive phenomena to the reader's most careful consideration, first, because they are among the strongest supports of the Theory of Descent, and secondly, because, considering their immense general importance, they have hitherto been properly considered only by a few privileged persons.

We cannot indeed but be astonished when we consider the deep ignorance which still prevails, in the widest circles, about the facts of the individual development of man and organisms in general. These facts, the universal importance of which cannot be estimated too highly, were established, in their most important outlines, even more than a hundred years ago, in 1759, by the great German naturalist Caspar Friedriech Wolff, in his classical "Theoria Generationis." But, just as Lamarck's Theory of Descent, founded in 1809, lay dormant for half a century, and was only awakened to new and imperishable life in 1859, by Darwin, in like manner Wolff's Theory of Epigenesis remained unknown for nearly half a century; and it was only after Oken, in 1806 had published his history of the development of the intestinal tube, and after Meckel, in 1812, had translated Wolff's work (written in Latin) on the same subject into German, that Wolff's theory of epigenesis became more generally known, and formed the foundation of all subsequent investigations of the history of individual development. The study of ontogenesis now received a great stimulus, and soon there appeared the classical investigations of the two friends, Christian Pander (1817) and Carl Ernst Bär (1819). Bär, in his remarkable "Entwickelungsgeschichte der Thiere,"[20] worked out the ontogeny of vertebrate animals in all its important facts. He carried out a series of such excellent observations, and illustrated them by such profound philosophical reflections, that his work became the foundation for a thorough understanding of this important group of animals, to which, of course, man also belongs. The facts of embryology alone would be sufficient to solve the question of man's position in nature, which is the highest of all problems. Look attentively at and compare the eight figures which are represented on the adjoining Plates II. and III., and it will be seen that the philosophical importance of embryology cannot be too highly estimated.

We may well ask, What do our so-called "educated" circles, who think so much of the high civilization of the 19th century, know of these most important biological facts, of these indispensable foundations for understanding their own organism? How much do our speculative philosophers and theologians know about them, who fancy they can arrive at an understanding of the human organism by mere guesswork or divine inspiration? What indeed do the majority of naturalists, not excepting the majority of the so-called "zoologists" (including the entomologists!), know about them?

The answer to this question tells much to the shame of the persons above indicated, and we must confess, willingly or unwillingly, that these invaluable facts of human ontogeny are, even at the present day, utterly unknown to most people, or are in no way valued as they deserve to be. It is in the face of such a condition of things as this that we see clearly upon what a wrong and one-sided road the much vaunted culture of the 19th century still moves. Ignorance and superstition are the foundations upon which most men construct their conception of their own organism and its relation to the totality of things; and these palpable facts of the history of development, which might throw the light of truth upon them, are ignored. It is true these facts are not calculated to excite approval among those who assume a thorough difference between man and the rest of nature, and who will not acknowledge the animal origin of the human race. That origin must be a very unpleasant truth to members of the ruling and privileged castes in those nations among which there exists an hereditary division of social classes, in consequence of false ideas about the laws of inheritance. It is well known that, even in our day, in many civilized countries the idea of hereditary grades of rank goes so far, that, for example, the aristocracy imagine themselves to be of a nature totally different from that of ordinary citizens, and nobles who commit a disgraceful offence are punished by being expelled from the caste of nobles, and thrust down among the pariahs of "vulgar citizens." What are these nobles to think of the noble blood which flows in their privileged veins, when they learn that all human embryos, those of nobles as well as commoners, during the first two months of development, are scarcely distinguishable from the tailed embryos of dogs and other mammals?

As the object of these pages is solely to further the general knowledge of natural truths, and to spread, in wider circles, a natural conception of the relations of man to the rest of nature, I shall be justified if I do not pay any regard to the widely-spread prejudice in favour of an exceptional and privileged position for man in creation, and simply give here the embryological facts from which the reader will be able to draw conclusions affirming the groundlessness of those prejudices. I wish all the more to entreat him to reflect carefully upon these facts of ontogeny, as it is my firm conviction that a general knowledge of them can only promote the intellectual advance, and thereby the mental perfecting, of the human race.

Amidst all the infinitely rich and interesting material which lies before us in the ontogeny of vertebrate animals, that is, in the history of their individual development, I shall here confine myself to showing some of those facts which are of the greatest importance to the Theory of Descent in general, as well as in its special application to man. Man is at the beginning of his individual existence a simple egg, a single little cell, just the same as every animal organism which originates by sexual generation. The human egg is essentially the same as that of all other mammals, and cannot be distinguished from the egg of the higher mammals. The egg represented in Fig. 5 might be that of a man or an ape as well as of a dog, a horse, or any other mammal. Not only the form and structure, but even the size of the egg in most mammals is the same as in man, namely, about the 120th part of an inch in diameter, so that the egg under favorable circumstances, with the naked eye, can just be perceived as a small speck. The differences which really exist between the eggs of different mammals and that of man do not consist in the form, but in the chemical mixture, in the molecular composition of the albuminous combination of carbon, of which the egg essentially consists. These minute individual differences of all eggs, which depend upon indirect or potential adaptation (and especially upon the law of individual adaptation), are indeed not directly perceptible to the exceedingly imperfect senses of man, but are cognisable through indirect means, as the primary causes of the difference of all individuals.

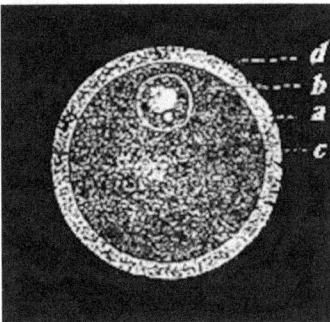

FIG. 5.—The human egg a hundred times enlarged. *a.* The kernel speck, or nucleolus (the so-called germinal spot of the egg). *b.* Kernel, or nucleus (the so-called germinal vesicle of the egg). *c.* Cell-substance, or protoplasm (so-called yolk of the egg). *d.* Cell-membrane (the yolk-membrane of the egg; in mammals, on account of its transparency, called zona pellucida). The eggs of other mammals are of the same form.

The human egg is, like that of all other mammals, a small globular bladder, which contains all the constituent parts of a simple organic cell (Fig. 5). The most essential parts of it are the mucous cell-substance, or the protoplasma (*c*), which in an egg is called the "yolk," and the cell-kernel, or nucleus (*b*), surrounded by it, which is here called by the special name of the "germinal vesicle." The latter is a delicate, clear, glassy globule of albumen, of about 1/600th part of an inch in diameter, and surrounds, a still smaller, sharply-marked, rounded granule (*a*), the kernel-speck, or the nucleolus of the cell (in the egg it is called the "germinal spot"). The outside of the globular egg-cell of a mammal is surrounded by a thick pellucid membrane, the cell-membrane or yolk-membrane, which here bears the special name of zona pellucida (*d*). The eggs of many lower animals (for example of many Medusae) differ from this in being *naked* cells, as the outer covering, or cell-membrane, is wanting.

As soon as the egg (ovulum) of the mammal has attained its full maturity, it leaves the ovary of the female, in which it originates, and passes into the oviduct, and through this narrow passage into the wider pouch or womb (uterus). If, meanwhile, the egg is fructified by the male seed (sperm), it develops itself in this pouch into an embryo, and does not leave it until perfectly developed and capable of coming into the world at birth as a young mammal.

The variations of form and transformations which the fructified egg must go through within the uterus before it assumes the form of the mammal are exceedingly remarkable, and proceed from the beginning in man, in precisely the same way as in the other mammals. At first the fructified egg of the mammal acts as a single-celled organism, which is about to propagate independently and increase itself; for example, an Amoeba (compare Fig. 2, p. 188 [108]). In point of fact the simple egg-cell becomes two, by the process of cell-division which I have previously described. There arise from the single germinal spot (the small kernel-speck of the original simple egg-cell) two new kernel-specks, and then in like manner, out of the germinal vesicle (the nucleus), two new cell-kernels. Then, and not until then, does the globular protoplasma first separate itself by an equatorial furrow into two halves, in such a manner that each half encloses one of the two kernels, together with its kernel-speck. Thus the simple egg-cell, within the original cellular membrane, has become two naked cells, each possessing its own kernel (Fig. 6).

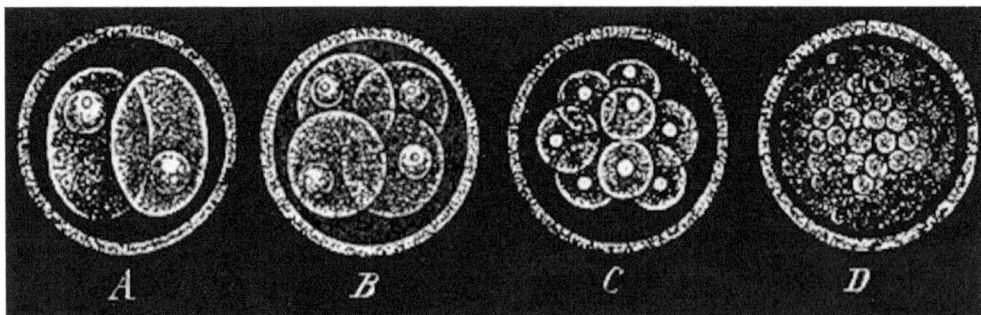

FIG. 6.—First commencement of the development of a mammal's egg, the so-called "yolk-cleavage" (propagation of the egg-cell by repeated self-division). *A*. The egg, by the formation of the first furrow, falls into two cells. *B*. These by division fall into four cells. *C*. These latter have fallen into eight cells. *D*. By continued division a globular mass of numerous cells has arisen.

The same process of cell-division now repeats itself several times in succession. In this way, from two cells (Fig. 6 *A*) there arise four (Fig. 6 *B*); from four, eight (Fig. 6 *C*); from eight, sixteen; from these, thirty-two, etc. Each time the division of the kernel-speck precedes that of the kernel; this, again, precedes that of the cell-substance, or protoplasma. As the division of the latter always commences with the formation of a superficial annular *furrow*, or cleft, the whole process is usually called the *furrowing of the egg*, or yolk-cleavage, and the products of it, that is, the cells arising from the continued halving, are called the *cleavage spheres*. However, the whole process is nothing more than a simple, oft-repeated *division of cells*, and the products of it are actual, naked *cells*. Finally, through the continued division or "furrowing" of the mammal's egg there arises a mulberry-shaped ball, which is composed of a great number of small spheres, naked cells, containing kernels (Fig. 6 *D*). These cells are the materials out of which the body of the young mammal is constructed. Every one of us has once been such a simple mulberry-shaped ball, composed only of small equi-formal cells.

The further development of the globular lump of cells, which now represents the young body of the mammal, consists first in its changing into a globular bladder, as fluid accumulates within it. This bladder is called the germ-bladder (vesicula blastodermica). Its wall is at first composed of merely equi-formal cells. But soon, at one point on the wall, arises a disc-shaped thickening, as the cells here increase rapidly, and this thickening is now the foundation of the actual body of the germ or embryo, while the other parts of the germ-bladder serve only for its nutrition. The thickened disc, or foundation of the embryo, soon assumes an oblong, and then a fiddle-shaped form, in consequence of its right and left walls becoming convex (Fig. 7, p. 304 [173]). At this stage of development in the first form of their germ or embryo, not only all mammals, including man, but even all vertebrate animals in general—birds, reptiles, amphibious animals, and fishes—can either not be distinguished from one another at all, or only by very unessential differences, such as the arrangement of the egg-coverings. In all the whole body consists of nothing but a quite simple, oblong, oval, or violin-shaped thin disc, which is composed of three closely connected membranes or plates, lying one above another. Each of the three plates or layers of the germ consists simply of cells all exactly like one another; but each layer has a different function in the building up of the vertebrate animal body. Out of the upper or outer germ-layer arises solely the outer skin (epidermis), together with the central parts of the nervous system (spinal marrow and brain); out of the lower or inner layer arises only the inner delicate skin (epithelium) which lines the whole intestinal tube from the mouth to the anus, together with all the glands connected with it (lung, liver, salivary glands, etc.); out of the middle germ-layer lying between the two others arise all the other organs, muscles, bones, blood-vessels. Now, the processes by which the various and exceedingly complicated parts of the fully-formed body of vertebrate animals arise out of such simple material—out of the three germ-layers composed only of cells—are, in the first place, the repeated division, and consequently the increase of cells; in the second place, the division of labour or differentiation of these cells; and thirdly, the union of the variously developed or differentiated cells, for the formation of the different organs. Thus arises the gradual progress or perfecting which can be traced step by step in the development of the embryonic body. The simple embryonic cells, which are to constitute the body of the vertebrate animal, stand in the same relation to each other as citizens who wish to found a state. Some take to one occupation, others to another, and work together for the good of the whole. By this division of labour, or differentiation, and the perfecting (the organic progress) which is connected with it, it becomes possible for the whole state to accomplish undertakings which would have been impossible to the single individual. The whole body of the vertebrate animal, like every other many-celled organism, is a republican state of cells, and consequently it can accomplish organic functions which the individual cell, as a solitary individual (for example, an Amoeba, or a single-celled plant), could never perform.

No sensible person supposes that carefully devised institutions, which have been established for the good of the whole, as well as for the individual, in every human state, are the results of the action of a personal and supernatural Creator, acting for a definite purpose. On the contrary, every one knows that these useful institutions of organization in the state are the consequences of the co-operation of the individual citizens and their common government, as well as of adaptation to the conditions of existence of the outer world. Just in the same way we must judge of the many-celled organism. In it also all the useful arrangements are solely the natural and necessary result of the co-operation, differentiation, and perfecting of the individual citizens—the cells—and by no means the artificial arrangements of a Creator acting for a definite purpose. If we rightly consider this comparison, and pursue it further, we can distinctly see the perversity of that dualistic conception of nature which discovers the action of a creative plan of construction in the various adaptations of the organization of living things. Let us pursue the individual development of the vertebrate animal body a few stages further, and see what is next done by the citizens of this embryonic organism. In the central line of the violin-shaped disc, which is composed of the three cellular germ-layers, there arises a straight delicate furrow, the so-called "primitive streak," by which the violin-shaped body is divided into two equal lateral halves—a right and a left part or "antimer." On both sides of that streak or furrow, the upper or external germ-layer rises in the form of a longitudinal fold, and both folds then grow together over the furrow in the central line, and thus form a cylindrical tube. This tube is called the marrow-tube, or medullary canal, because it is the foundation of the central nervous system, the *spinal marrow* (medulla spinalis). At first it is pointed both in front and behind, and it remains so for life in the lowest vertebrate animal, the brainless, skull-less Lancelet (*Amphioxus*). But in all other vertebrate animals, which we distinguish from the latter as skulled animals, or Craniota, a difference between the fore and hinder end of the marrow tube soon becomes visible, the fore end becoming dilated, and changing into a roundish bladder, the foundation of the *brain*.

In all Craniota, that is, in all vertebrate animals possessing skull and brain, the brain, which is at first only the bladder-shaped dilatation of the anterior end of the spinal marrow, divides into five bladders lying one behind the other, four superficial, transverse in-nippings being formed. These *five brain-bladders*, out of which afterwards arise all the different parts of the intricately constructed brain, can be seen in their original condition in the embryo represented in Fig. 7. It is just the same whether we examine the embryo of a dog, a fowl, a lizard, or any other higher vertebrate animal. For the embryos of the different skulled animals (at least the three higher classes of them, the reptiles, birds and mammals) cannot be in any way distinguished at the stage represented in Fig. 7. The whole form of the body is as yet exceedingly simple, being merely a thin, leaf-like disc. Face, legs, intestines, etc., are as yet completely wanting. But the five bladders are already quite distinct from one another.

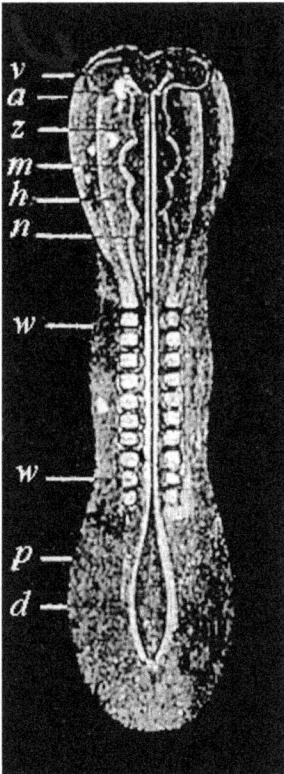

FIG. 7.—Embryo of a mammal or bird, in which the five brain-bladders have just commenced to develop. *v*. Fore brain. *z*. Twixt brain. *m*. Mid brain. *h*. Hind brain. *n*. After brain. *p*. Spinal-marrow. *a*. Eye-bladders. *w*. Primitive vertebrae. *d*. Spinal-axis or notochord.

The *first* bladder, the *fore brain* (*a*), is in so far the most important that it principally forms the hemispheres of the so-called larger brain (cerebrum), that part which is the seat of the higher mental activities. The more these activities are developed in the series of vertebrate animals, the more do the two lateral halves of the fore brain, or the hemispheres, grow at the expense of the other bladders, and overlap them in front and from above. In man, where they are most strongly developed, agreeing with his higher mental activity, they eventually almost entirely cover the other parts from above (compare Plates II. and III.) The *second* bladder, the *twixt brain* (*z*), forms that portion of the brain which is called the *centre of sight*, and stands in the closest relation to the eyes (*a*), which grow right and left out of the fore brain in the shape of two bladders, and later lie at the bottom of the twixt brain. The *third* bladder, the *mid brain* (*m*), for the most part vanishes in the formation of the so-called *four bulbs*, a bossy portion of the brain, which is strongly developed in reptiles and birds (Fig. *E*, *F*, Plate II.), whereas in mammals it recedes much more (Fig. *G*, *H*, Plate III.). The *fourth* bladder, the *hind brain* (*h*), forms the so-called *little hemispheres*, together with the middle part of the *small brain* (cerebellum), a part of the brain as to the function of which the most contradictory conjectures are entertained, but which seems principally to regulate the co-ordination of movements. Lastly, the *fifth* bladder, the *after brain* (*n*), develops into that very important part of the central nervous system which is called the *prolonged marrow* (medulla oblongata). It is the central organ of the respiratory movements, and of other important functions, and an injury to it immediately causes death, whereas the large hemispheres of the fore brain (or the organ of the "soul," in a restricted sense) can be removed bit by bit, and even completely destroyed, without causing the death of the vertebrate animal—only its higher mental activities disappearing in consequence.

These five brain bladders, in all vertebrate animals which possess a brain at all, are originally arranged in the same manner and develop gradually in the different groups so differently, that it is afterwards very difficult to recognize the corresponding parts in the fully-developed brains. In the early stage of development which is represented in Fig. 7, it seems as yet quite impossible to distinguish the embryos of the different mammals, birds, and reptiles, from one another. But if we compare the much more developed embryos on Plates II. and III. with one another, we can clearly see an inequality in their development, and especially it will be perceived that the brain of the two mammals (*G* and *H*) already strongly differ from that of birds (*F*) and of reptiles (*E*). In the two latter the mid brain predominates, but in the former the fore brain. Even at this stage the brain of the bird (*F*) is scarcely distinguishable from that of the tortoise (*E*), and in like manner the brain of the dog (*G*) is as yet almost the same as that of man (*H*). If, on the other hand, we compare the brains of these four vertebrate animals in a fully developed condition, we find them so very different in all anatomical particulars, that we cannot doubt for a moment as to which animal each brain belongs.

I have here explained the original equality, the gradual commencement, and the ever increasing separation or differentiation of the embryos in the different vertebrate animals, taking the brain as a special example, just because this organ of the soul's activity is of special interest. But I might as well have discussed in its stead the heart, or the liver, or the limbs, in short, any other part of the body, since the same wonder of creation is here ever repeated, namely, this, that all parts are originally the same in the different vertebrate animals, and that the variations by which the different classes, orders, families, genera, etc., differ and deviate from one another, are only gradually developed.

There are certainly few parts of the body which are so differently constructed as the *limbs or extremities* of the vertebrate animals. Now, I wish the reader to compare in Fig. *A—H* on Plates II. and III., the four extremities (*bv*) of the embryos with one another, and he will scarcely be able to perceive any important differences between the human arm (*H bv*), the wing of a bird (*F bv*), the slim foreleg of a dog (*G bv*), and the plump foreleg of the tortoise (*E bv*). In comparing the hinder extremities (*bh*) in these figures he will find it equally difficult to distinguish the leg of a man (*H bh*), of a bird (*F bh*), the hind-leg of a dog (*G bh*), and that of a tortoise (*E bh*). The fore as well as the hinder extremities are as yet short, broad lumps, at the ends of which the foundations of the five toes are placed, connected as yet by a membrane. At a still earlier stage (Fig. *A—D*) the five toes are not marked out at all, and it is quite impossible to distinguish even the fore and hinder extremities from one another. The latter, as well as the former, are nothing but simple roundish processes, which have grown out of the side of the trunk. At the very early stage represented in Fig. 7 they are completely wanting, and the whole embryo is a simple trunk without a trace of limbs.

I wish especially to draw attention in Plates II. and III., which represents embryos in early stages of development (Fig. *A—D*)—and in which we are not able to recognize a trace of the full-grown animal—to an exceedingly important formation, which originally is common to all vertebrate animals, but which at a later period is transformed into the most different organs. Every one surely knows the *gill-arches* of fish, those arched bones which lie behind one another, to the number of three or four, on each side of the neck, and which support the gills, the respiratory organs of the fish (double rows of red leaves, which are popularly called "fishes' ears."). Now, these gill-arches originally exist exactly the same in man (*D*), in dogs (*C*), in fowls (*B*), and in tortoises (*A*), as well as in all other vertebrate animals. (In Fig. *A—D* the three gill-arches of the right side of the neck are marked k_1 k_2 k_3). Now, it is only in fishes that these remain in their original form, and develop into respiratory organs. In the other vertebrate animals they are partly employed in the formation of the face (especially the jaw apparatus), and partly in the formation of the organ of hearing.

Finally, when comparing the embryos on Plates II. and III., we must not fail to give attention again to the *human tail* (*s*), an organ which, in the original condition, man shares with all other vertebrate animals. The discovery of tailed men was long anxiously expected by many monistic philosophers, in order to establish a closer relationship between man and the other mammals. And in like manner their dualistic opponents often maintained with pride that the complete want of a tail formed one of the most important bodily distinctions between men and animals, though they did not bear in mind the many tailless animals which really exist. Now, man in the first months of development possesses a real tail as well as his nearest kindred, the tailless apes (orang-outang, chimpanzee, gorilla), and vertebrate animals in general. But whereas, in most of them—for example, the dog (*C*, *G*)—in the course of development it always grows longer, in man (Fig. *D*, *H*)

Pl. II.   Germs or Embryos

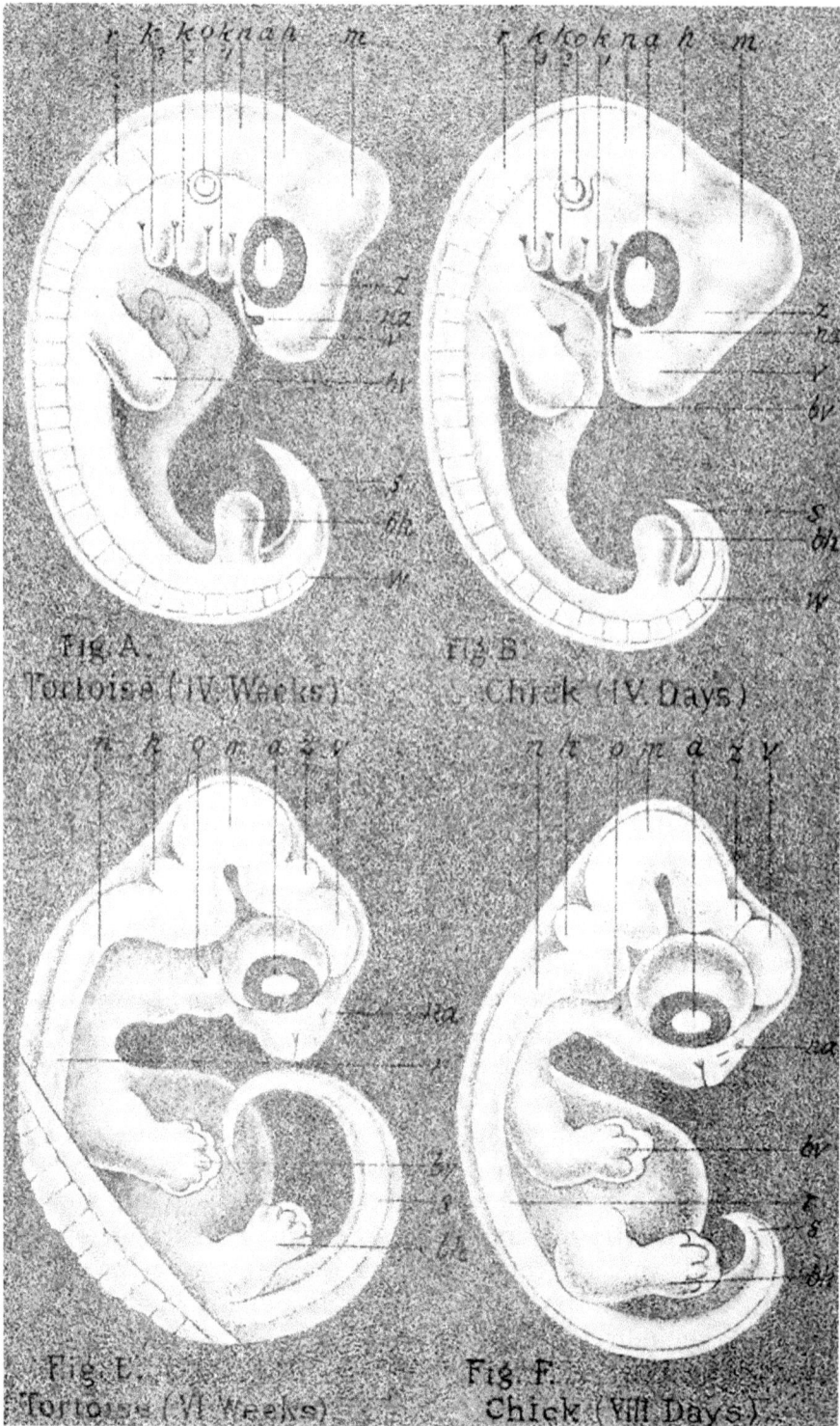

Fig. A.
Tortoise (IV Weeks)

Fig. B.
Chick (IV Days)

Fig. E.
Tortoise (VI Weeks)

Fig. F.
Chick (VIII Days)

v. Fore-brain. *z.* Twixt-brain. *m.* Mid-brain. *h.* Hind-brain.
n. After-brain. *w.* Spine. *r.* Spinal-cord.

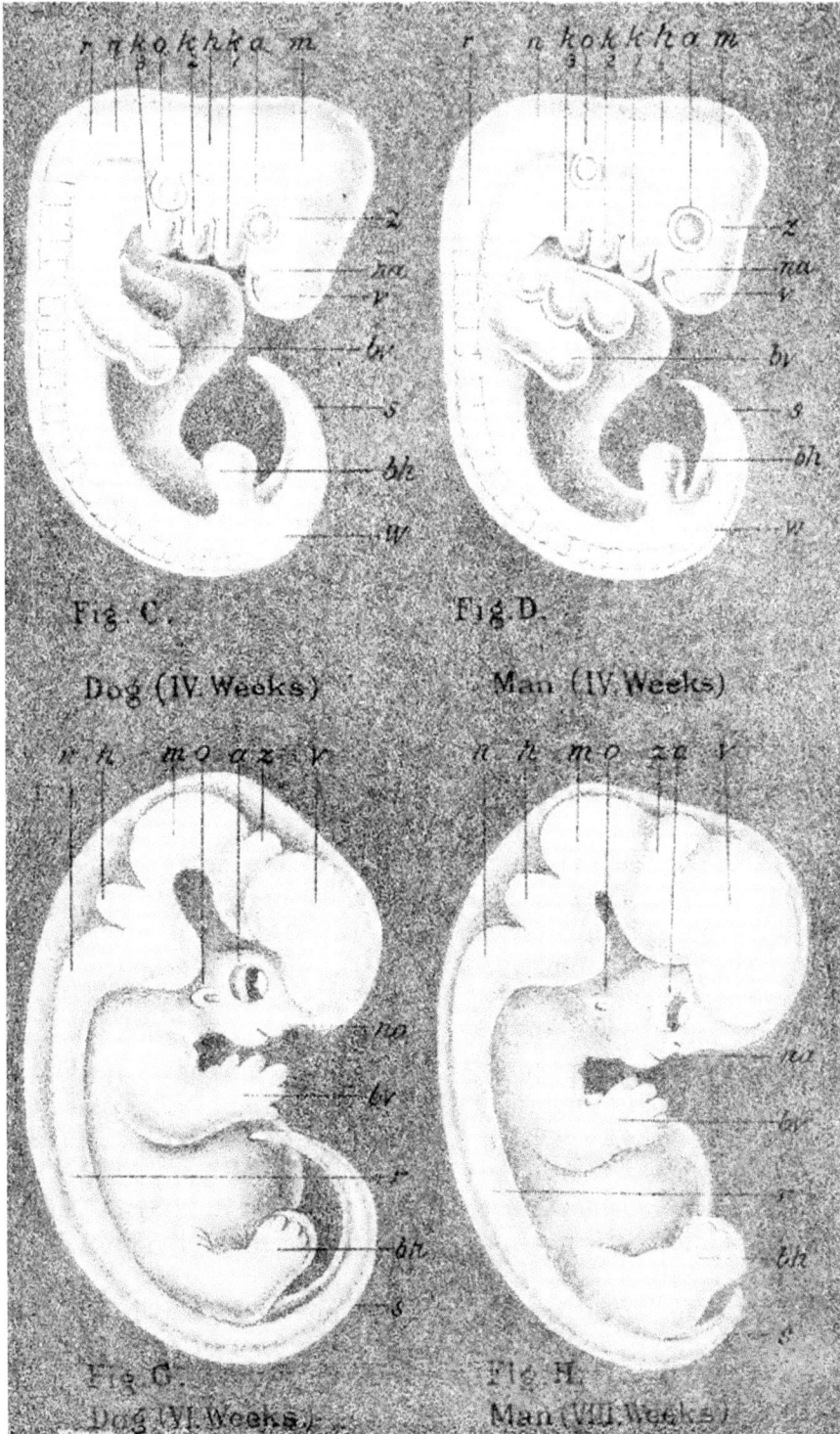

Fig. C.

Dog (IV. Weeks)

Fig. D.

Man (IV. Weeks)

Fig. G.

Dog (VI. Weeks)

Fig. H.

Man (VIII. Weeks)

na. Nose. *a*. Eyes. *o*. Ear. k_1 k_2 k_3 Gill-arches. *s*. Tail. *bv*. Fore-leg. *bh*. Hind-leg.

and in tailless mammals, at a certain period of development, it degenerates and finally completely disappears. However, even in fully developed men, the remnant of the tail is seen in the three, four, or five tail vertebrae (vertebrae coccygeae) as an aborted or rudimentary organ, which forms the hinder or lower end of the vertebral column (p. 289 [165]).

Most persons even now refuse to acknowledge the most important deduction of the Theory of Descent, that is, the palaeontological development of man from ape-like, and through them from still lower, mammals, and consider such a transformation of organic form as impossible. But, I ask, are the phenomena of the individual development of man, the fundamental features of which I have here given, in any way less wonderful? Is it not in the highest degree remarkable that all vertebrate animals of the most different classes—fishes, amphibious animals, reptiles, birds, and mammals—in the first periods of their embryonic development cannot be distinguished at all, and even much later, at a time when reptiles and birds are already distinctly different from mammals, that the dog and the man are almost identical? Verily, if we compare those two series of development with one another, and ask ourselves which of the two is the more wonderful, it must be confessed that *ontogeny*, or the short and quick history of development of the *individual*, is much more mysterious than *phylogeny*, or the long and slow history of development of the *tribe*. For one and the same grand change of form is accomplished by the latter in the course of many thousands of years, and by the former in the course of a few months. Evidently this most rapid and astonishing transformation of the individual in ontogenesis, which we can actually point out at any moment by direct observation, is in itself much more wonderful and astonishing than the corresponding, but much slower and gradual transformation which the long chain of ancestors of the same individual has gone through in phylogenesis.

The two series of organic development, the ontogenesis of the individual and the phylogenesis of the tribe to which it belongs, stand in the closest causal connection with each other. I have endeavoured, in the second volume of the "General Morphology,"[4] to establish this theory in detail, as I consider it exceedingly important. As I have there shown, *ontogenesis, or the development of the individual, is a short and quick repetition* (recapitulation) *of phylogenesis, or the development of the tribe to which it belongs, determined by the laws of inheritance and adaptation*; by tribe I mean the ancestors which form the chain of progenitors of the individual concerned. (Gen. Morph. ii. 110–147, 371.)

In this intimate connection of ontogeny and phylogeny, I see one of the most important and irrefutable proofs of the Theory of Descent. No one can explain these phenomena unless he has recourse to the laws of Inheritance and Adaptation; by these alone are they explicable. These laws, which we have previously explained, are *the laws of abbreviated, of homochronic, and of homotopic inheritance*, and here deserve renewed consideration. As so high and complicated an organism as that of man, or the organism of every other mammal, rises upwards from a simple cellular state, and as it progresses in its differentiation and perfecting it passes through the same series of transformations which its animal progenitors have passed through, during immense spaces of time, inconceivable ages ago. I have already pointed out this extremely important parallelism of the development of individuals and tribes (p. 10 [6]). Certain very early and low stages in the development of man, and the other vertebrate animals in general, correspond completely in many points of structure with conditions which last for life in the lower fishes. The next phase which follows upon this presents us with a change of the fish-like being into a kind of amphibious animal. At a later period the mammal, with its special characteristics, develops out of the amphibian, and we can clearly see, in the successive stages of its later development, a series of steps of progressive transformation which evidently correspond with the differences of different mammalian orders and families. Now, it is precisely in the same succession that we also see the ancestors of man, and of the higher mammals, appear one after the other in the earth's history; first fishes, then amphibians, later the lower, and at last the higher mammals. Here, therefore, the embryonic development of the individual is completely parallel to the palaeontological development of the whole tribe to which it belongs, and this exceedingly interesting and important phenomenon can be explained only by the interaction of the laws of Inheritance and Adaptation.

The example last mentioned, of the parallelism of the palaeontological and of the individual developmental series, now directs our attention to a third developmental series, which stands in the closest relations to these two, and which likewise runs, on the whole, parallel to them. I mean that series of development of forms which constitutes the object of investigation in *comparative anatomy*, and which I will briefly call the *systematic developmental series of species*. By this we understand the chain of the different, but related and connected forms, which exist *side by side* at any one period of the earth's history; as for example, at the present moment. While comparative anatomy compares the different forms of fully-developed organisms with one another, it endeavours to discover the common prototypes which underlie, as it were, the manifold forms of kindred genera, classes, etc., and which are more or less concealed by their particular differentiation. It endeavours to make out the series of progressive steps which are indicated in the different degrees of perfection of the divergent branches of the tribe. To make use again of the same particular instance, comparative anatomy shows us how the individual organs and systems of organs in the tribe of vertebrate animals—in the different classes, families, and species of it—have unequally developed, differentiated, and perfected themselves. It shows us how far the succession of classes of vertebrate animals, from the Fishes upwards, through the Amphibia to the Mammals, and here again, from the lower to the higher orders of Mammals, forms a progressive series or ladder. This attempt to establish a connected anatomical developmental series we may discover in the works of the great comparative anatomists of all ages—in the works of Goethe, Meckel, Cuvier, Johannes Müller, Gegenbaur, and Huxley.

The developmental series of mature forms, which comparative anatomy points out in the different diverging and ascending steps of the organic system, and which we call the systematic developmental series, is parallel to the palaeontological developmental series, because it deals with the *result* of palaeontological development, and it is parallel to the individual developmental series, because this is parallel to the palaeontological series. If two parallels are parallel to a third, they must be parallel to one another.

The varied differentiation, and the unequal degree of perfecting which comparative anatomy points out in the developmental series of the System, is chiefly determined by the ever increasing variety of conditions of existence to which the different groups adapt themselves in the struggle for life, and by the different degrees of rapidity and completeness with which this adaptation has been effected. Conservative groups which have retained their inherited peculiarities most tenaciously remain, in consequence, at the lowest and rudest stage of development. Those groups progressing most rapidly and variously, and which have adapted themselves to changed conditions of existence most readily have attained the highest degree of perfection. The further the organic world developed in the course of the earth's history, the greater must the gap between the lower conservative and the higher progressive groups have become, as in fact may be seen too in the history of nations. In this way also is explained the historical fact, that the most perfect animal and vegetable groups have developed themselves in a comparatively short time to a considerable height, while the lowest or most conservative groups have remained stationary throughout all ages in their original simple stage, or have progressed, but very slowly and gradually. The series of man's progenitors clearly shows this state of things. The sharks of the present day are still very like the primary fish, which are among the most ancient vertebrate progenitors of man, and the lowest amphibians of the present day (the gilled salamanders and salamanders) are very like the amphibians which first developed themselves out of fishes. So, too, the later ancestors of man, the Monotremata and Marsupials, the most ancient mammals, are at the same time the most imperfect animals of the class which still exist.

The laws of inheritance and adaptation known to us are completely sufficient to explain this exceedingly important and interesting phenomenon, which may be briefly designated as the *parallelism of individual, of palaeontological, and of systematic development.* No opponent of the Theory of Descent has been able to give an explanation of this extremely wonderful fact, whereas it is perfectly explained, according to the Theory of Descent, by the laws of Inheritance and Adaptation.

If we examine this parallelism of the three organic series of development more accurately, we have to add the following special qualifications. *Ontogeny*, or the history of the individual development of every organism (embryology and metamorphology), presents us with a simple *unbranching* or graduated chain of forms; and so it is with that *portion of phylogeny* which comprises the palaeontological history of development of the *direct ancestors only* of an individual organism. But *the whole of phylogeny*—which meets us in the *natural system* of every organic tribe or phylum, and which is concerned with the investigation of the palaeontological development *of all* the branches of this tribe—forms a *branching* or tree-shaped developmental series, a veritable pedigree. If we examine and compare the branches of this pedigree, and place them together according to the degree of their differentiation and perfection, we obtain the tree-shaped, branching, *systematic developmental series of comparative anatomy*. Strictly speaking, therefore, the latter is parallel to *the whole of phylogeny*, and consequently is only partially parallel to ontogeny; for ontogeny itself is parallel only to *a portion* of phylogeny.

All the phenomena of organic development above discussed, especially the threefold genealogical parallelism, and the laws of differentiation and progress, which are evident in each of these three series of organic development, and, further, the whole history of rudimentary organs, are exceedingly important proofs of the truth of the Theory of Descent. For by it alone can they be explained, whereas its opponents cannot even offer a shadow of an explanation of them. Without the Doctrine of Filiation, the fact of organic development in general cannot be understood. We should therefore, for this reason alone, be forced to accept Lamarck's Theory of Descent, even if we did not possess Darwin's Theory of Selection.

CHAPTER XIII.

THEORY OF THE DEVELOPMENT OF THE UNIVERSE AND OF THE EARTH. SPONTANEOUS GENERATION. THE CARBON THEORY. THE PLASTID THEORY.

History of the Development of the Earth.—Kant's Theory of the Development of the Universe, or the Cosmological Gas Theory.—Development of Suns, Planets, and Moons.—First Origin of Water.—Comparison of Organisms and Anorgana.—Organic and Inorganic Substances.—Degrees of Density, or Conditions of Aggregation.—Albuminous Combinations of Carbon.—Organic and Inorganic Forms.—Crystals and Formless Organisms without Organs.—Stereometrical Fundamental Forms of Crystals and of Organisms.—Organic and Inorganic Forces.—Vital Force.—Growth and Adaptation in Crystals and in Organisms.—Formative Tendencies of Crystals.—Unity of Organic and Inorganic Nature.—Spontaneous Generation, or Archigony.—Autogony and Plasmogony.—Origin of Monera by Spontaneous Generation.—Origin of Cells from Monera.—The Cell Theory.—The Plastid Theory.—Plastids, or Structural Units.—Cytods and Cells.—Four Different Kinds of Plastids.

IN our considerations hitherto we have endeavoured to answer the question, "By what causes have new species of animals and plants arisen out of existing species?" We have answered this question according to Darwin's theory, that natural selection in the struggle for existence—that is, the interaction of the laws of Inheritance and Adaptation—is completely sufficient for producing mechanically the endless variety of the different animals and plants, which have the appearance of being organized according to a plan for a definite purpose. Meanwhile the question must have already repeatedly presented itself to the reader, how did the first organisms, or that one original and primaeval organism arise, from which we derive all the others?

This question Lamarck[2] answered by the hypothesis of *spontaneous generation*, or *archigony*. But Darwin passes over and avoids this subject, as he expressly remarks that he has "nothing to do with the origin of the soul, nor with that of life itself." At the conclusion of his work he expresses himself more distinctly in the following words—"I imagine that probably all organic beings which ever lived on this earth descended from some primitive form, which was first called into life by the Creator." Moreover, Darwin, for the consolation of those who see in the Theory of Descent the destruction of the whole "moral order of the universe," appeals to the celebrated author and divine who wrote to him, that "he has gradually learnt to see that it is just as noble a conception of the Deity to believe that he created a few original forms capable of self-development into other and needful forms, as to believe that he required a fresh act of creation to supply the voids caused by the action of his laws."

Those to whom the belief in a supernatural creation is an emotional necessity may rest satisfied with this conception. They may reconcile that belief with the Theory of Descent; for in the creation of a single original organism possessing the capability to develop all others out of itself by inheritance and adaptation, they can really find much more cause for admiring the power and wisdom of the Creator than in the independent creation of different species.

If, taking this point of view, we were to explain the origin of the first terrestrial organisms, from which all the others are descended, as due to the action of a personal Creator acting according to a definite plan, we should of course have to renounce all scientific knowledge of the process, and pass from the domain of true science to the completely distinct domain of poetical faith. By assuming a supernatural act of creation we should be taking a leap into the inconceivable. Before we decide upon this latter step, and thereby renounce all pretension to a scientific knowledge of the process, we are at all events in duty bound to endeavour to examine it in the light of a mechanical hypothesis. We must at least examine whether this process is really so wonderful, and whether we cannot form a tenable conception of a completely non-miraculous origin of the first primary organism. We might then be able entirely to reject miracle in creation.

It will be necessary for this purpose, first of all, to go back further into the past, and to examine the history of the creation of the earth. Going back still further, we shall find it necessary to consider the history of the creation of the whole universe in its most general outlines. All my readers undoubtedly know that from the structure of the earth, as it is at present known to us, the notion has been derived, and as yet has not been refuted, that its interior is in a fiery fluid condition, and that the firm crust, composed of different strata, on the surface of which organisms are living, forms only a very thin pellicle or shell round the fiery fluid centre. We have arrived at this idea by different confirmatory experiments and reasonings. In the first place, the observation that the temperature of the earth's crust continually increases towards the centre is in favour of this supposition. The deeper we descend, the greater the warmth of the ground, and in such proportion, that with every 100 feet the temperature increases about one degree. At a depth of six miles, therefore, a heat of 1500 deg. would be attained, sufficient to keep most of the firm substances of our earth's crust in a molten, fiery, fluid state. This depth, however, is only the 286th part of the whole diameter of the earth (1717 miles). We further know that springs which rise out of a considerable depth possess a very high temperature, and sometimes even throw water up to the surface in a boiling state. Lastly, very important proofs are furnished by volcanic phenomena, the eruption of fiery fluid masses of stone bursting through certain parts of the earth's crust. All these phenomena lead us with great certainty to the important assumption that the firm crust of the earth forms only quite a small fraction, not nearly the one-thousandth part of the whole diameter of the terrestrial globe, and that the rest is still for the most part in a molten or fiery fluid state.

Now if, starting with this assumption, we reflect on the ancient history of the development of the globe, we are logically carried back a step further, namely, to the assumption that at an earlier date the whole earth was a fiery fluid body, and that the formation of a thin, stiffened crust on the surface was only a later process. Only gradually, by radiating its intrinsic heat into the cold space of the universe, has the surface of the glowing ball become condensed into a thin crust.

That the temperature of the earth in remote times was much higher than it is now, is proved by many phenomena. Among other things, this is rendered probable by the equal distribution of organisms in remote times of the earth's history. While at present, as is well known, the different populations of animals and plants correspond to the different zones of the earth and their appropriate temperature, in earlier times this was distinctly not the case.

We see from the distribution of fossils in the remoter ages, that it was only at a very late date, in fact, at a comparatively recent period of the organic history of the earth (at the beginning of the so-called caenolithic or tertiary period), that a separation of zones and of the corresponding organic populations occurred. During the immensely long primary and secondary periods, tropical plants, which require a very high degree of temperature, lived not only in the present torrid zone, under the equator, but also in the present temperate and frigid zones. Many other phenomena also demonstrate a gradual decrease of the temperature of the globe as a whole, and especially a late and gradual cooling of the earth's crust about the poles. Bronn, in his excellent "Investigations of the Laws of Development of the Organic World," has collected numerous geological and palaeontological proofs of this fact.

These phenomena and the mathematico-astronomical knowledge of the structure of the universe justify the theory that, inconceivable ages ago, long before the first existence of organisms, the whole earth was a fiery fluid globe. Now, this theory corresponds with the grand theory of the origin of the universe, and especially of our planetary system, which, on the ground of mathematical and astronomical facts, was put forward in 1755 by our critical philosopher Kant,[22] and was later more thoroughly established by the celebrated mathematicians, Laplace and Herschel. This cosmogeny, or theory of the development of the universe, is now almost universally acknowledged; it has not been replaced by a better one, and mathematicians, astronomers, and geologists have continually, by various arguments, strengthened its position.

Kant's cosmogeny maintains that *the whole universe, inconceivable ages ago, consisted of a gaseous chaos.* All the substances which are found at present separated on the earth, and other bodies of the universe, in different conditions of density—in the solid, semi-fluid, liquid, and elastic fluid or gaseous states of aggregation—originally constituted together one single homogeneous mass, equally filling up the space of the universe, which, in consequence of an extremely high degree of temperature, was in an exceedingly thin gaseous or nebulous state. The millions of bodies in the universe which at present form the different solar systems did not then exist. They originated only in consequence of a universal rotatory movement, or rotation, during which a number of masses acquired greater density than the remaining gaseous mass, and then acted upon the latter as central points of attraction. Thus arose a separation of the chaotic primary nebula, or gaseous universe, into a number of rotating nebulous spheres, which became more and more condensed. Our solar system was such a gigantic gaseous or nebulous ball, all the particles of which revolved round a common central point, the solar nucleus. The nebulous ball itself, like all the rest, in consequence of its rotatory movement, assumed a spheroidal or a flattened globular form.

While the centripetal force attracted the rotating particles nearer and nearer to the firm central point of the nebulous ball, and thus condensed the latter more and more, the centrifugal force, on the other hand, always tended to separate the peripheral particles further and further from it, and to hurl them off. On the equatorial sides of the ball, which was flattened at both poles, this centrifugal force was strongest, and as soon as, by increase of density, it attained predominance over the centripetal force, a circular nebulous ring separated itself from the rotating ball. This nebulous ring marked the course of future planets. The nebulous mass of the ring gradually condensed, and became a planet, which revolved round its own axis, and at the same time rotated round the central body. In precisely the same manner, from the equator of the planetary mass, as soon as the centrifugal force gained predominance over the centripetal force, new nebulous rings were ejected, which moved round the planets as the latter moved round the sun. These nebulous rings, too, became condensed into rotating balls. Thus arose the moons, only one of which moves round our earth, whilst four move round Jupiter, and six round Uranus. The ring of Saturn still shows us a moon in its early stage of development. As by increasing refrigeration these simple processes of condensation and expulsion repeated themselves over and over again, there arose the different solar systems, the planets rotating round their central suns, and the satellites or moons moving round their planets.

The original gaseous condition of the rotating bodies of the universe gradually changed, by increasing refrigeration and condensation, into the fiery fluid or molten state of aggregation. By the process of condensation, a great quantity of heat was emitted, and the rotating suns, planets, and moons, soon changed into glowing balls of fire, like gigantic drops of melted metal, which emitted light and heat. By loss of heat, the melted mass on the surface of the fiery fluid ball became further condensed, and thus arose a thin, firm crust, which enclosed a fiery fluid nucleus. In all essential respects our mother earth probably did not differ from the other bodies of the universe.

In view of the object of these pages, it will not be of especial interest to follow in detail the *history of the natural creation of the universe*, with its different solar and planetary systems, and to establish it mathematically by the different astronomical and geological proofs. The outlines of it, which I have just mentioned, must be sufficient here, and for further details I refer to Kant's* "General History of Nature and Theory of the Heavens."[22] I will only add that this wonderful theory, which might be called *the cosmological gas theory*, harmonizes with all the general series of phenomena at present known to us, and stands in no irreconcilable contradiction to any one of them. Moreover, it is purely mechanical or monistic, makes use exclusively of the inherent forces of eternal matter, and entirely excludes every supernatural process, every prearranged and conscious action of a personal Creator. Kant's Cosmological Gas Theory consequently occupies a similar supreme position in *Anorganology*, especially in *Geology*, and forms the crown of our knowledge in that department, in the same way as Lamarck's Theory of Descent does in *Biology*, and especially in *Anthropology*. Both rest exclusively upon mechanical or unconscious causes (causae efficientes), in no case upon prearranged or conscious causes (causae finales). (Compare above, p. 100–106. [57–60]) Both therefore fulfil all the demands of a scientific theory, and consequently will remain generally acknowledged until they are replaced by better ones.

I will, however, not deny that Kant's grand cosmogeny has some weak points, which prevent our placing the same unconditional confidence in it as in Lamarck's Theory of Descent. The notion of an original gaseous chaos filling the whole universe presents great difficulties of various kinds. A great and unsolved difficulty lies in the fact that the Cosmological Gas Theory furnishes no starting-point at all in explanation of the first impulse which caused the rotary motion in the gas-filled universe. In seeking for such an impulse, we are involuntarily led to the mistaken questioning about a "first beginning." We can as little imagine a *first beginning* of the eternal phenomena of the motion of the universe as of its final end.

The universe is unlimited and immeasurable in both space and time. It is eternal, and it is infinite. Nor can we imagine a beginning or end to the

---

* "Allgemeine Naturgeschichte und Theorie des Himmels."

uninterrupted and eternal motion in which all particles of the universe are always engaged. The great laws of the *conservation of force*[38] and the *conservation of matter*, the foundations of our whole conception of nature, admit of no other supposition. The universe, as far as it is cognisable to human capability, appears as a connected chain of material phenomena of motion, necessitating a continual change of forms. Every form, as the temporary result of a multiplicity of phenomena of motion, is as such perishable, and of limited duration. But, in the continual change of forms, matter and the motion inseparable from it remain eternal and indestructible.

Now, although Kant's Cosmological Gas Theory is not able to explain the development of motion in the whole universe in a satisfactory manner, beyond that gaseous state of chaos, and although many other weighty considerations may be brought forward against it, especially by chemistry and geology, yet we must on the whole acknowledge its great merit, inasmuch as it explains in an excellent manner, by due consideration of development, the whole structure of all that is accessible to our observation, that is, the anatomy of the solar systems, and especially of our planetary system. It may be that this development was altogether different from what Kant supposes, and our earth may have arisen by the aggregation of numberless small meteorides, scattered in space, or in any other manner, but hitherto no one has as yet been able to establish any other theory of development, or to offer one in the place of Kant's cosmogeny.

After this general glance at the monistic cosmogeny, or the non-miraculous history of the development of the universe, let us now return to a minute fraction of it, to our mother earth, which we left as a ball flattened at both poles and in a fiery fluid state, its surface having condensed by becoming cooled into a very thin firm crust. The crust, on first cooling, must have covered the whole surface of the terrestrial sphere as a continuous smooth and thin shell. But soon it must have become uneven and hummocky; for, since during the continued cooling, the fiery fluid nucleus became more and more condensed and contracted, and consequently the diameter of the earth diminished, the thin cold crust, which could not closely follow the softer nuclear mass, must have fallen in, in many places. An empty space would have arisen between the two, had not the pressure of the outer atmosphere forced down the fragile crust towards the interior, breaking it in so doing. Other unevennesses probably arose from the fact that, in different parts, the cooled crust during the process of refrigeration contracted also itself, and thus became fissured with cracks and rents. The fiery fluid nucleus flowed up to the external surface through these cracks, and again became cooled and stiff. Thus, even at an early period there arose many elevations and depressions, which were the first foundations of mountains and valleys.

After the temperature of the cooled terrestrial ball had fallen to a certain degree, a very important new process was effected, namely, the *first origin of water*. Water had until then existed only in the form of steam in the atmosphere surrounding the globe. The water could evidently not condense into a state of fluid drops until the temperature of the atmosphere had considerably decreased. Now, then, there began a further transformation of the earth's crust by the force of water. It continually fell in the form of rain, and in that form washed down the elevations of the earth's crust, filling the depressions with the mud carried along, and, by depositing it in layers, it caused the extremely important neptunic transformations of the earth's crust, which have continued since then uninterruptedly, and which in our next chapter we shall examine a little more closely.

It was not till the earth's crust had so far cooled that the water had condensed into a fluid form, it was not till the hitherto dry crust of the earth had for the first time become covered with liquid water, that the origin of the first organisms could take place. For all animals and all plants—in fact, all organisms—consist in great measure of fluid water, which combines in a peculiar manner with other substances, and brings them into a semi-fluid state of aggregation. We can therefore, from these general outlines of the inorganic history of the earth's crust, deduce the important fact, that at a certain definite time life had its beginning on earth, and that terrestrial organisms did not exist from eternity, but at a certain period came into existence for the first time.

Now, how are we to conceive of this origin of the first organisms? This is the point at which most naturalists, even at the present day, are inclined to give up the attempt at natural explanation, and take refuge in the miracle of an inconceivable creation. In doing so, as has already been remarked, they quit the domain of scientific knowledge, and renounce all further insight into the eternal laws which have determined nature's history. But before despondingly taking such a step, and before we despair of the possibility of any knowledge of this important process, we may at least make an attempt to understand it. Let us see if in reality the origin of a first organism out of inorganic matter, the origin of a living body out of lifeless matter, is so utterly inconceivable and beyond all experience. In one word, let us examine the question of *spontaneous generation, or archigony.* In so doing, it is above all things necessary to form a clear idea of the principal properties of the two chief groups of natural bodies, the so-called inanimate or inorganic, and the animate or organic bodies, and then establish what is common to, and what are the differences between, the two groups. It is desirable to go somewhat carefully into the *comparison of organisms and anorgana,* since it is commonly very much neglected, although it is necessary for a right understanding of nature from the monistic point of view. It will be most advantageous here to look separately at the three fundamental properties of every natural body; these are matter, form, and force. Let us begin with *matter.* (Gen. Morph. iii.)

By chemistry we have succeeded in analysing all bodies known to us into a small number of elements or simple substances, which cannot be further divided, for example, carbon, oxygen, nitrogen, sulphur, and the different metals: potassium, sodium, iron, gold, etc. At present we know about seventy such elements or simple substances. The majority of them are unimportant and rare; the minority only are widely distributed, and compose not only most of the anorgana, but also all organisms. If we compare those elements which constitute the body of organisms with those which are met with in anorgana, we have first to note the highly important fact that in animal and vegetable bodies no element occurs but what can be found outside of them in inanimate nature. There are no special organic elements or simple organic substances.

The chemical and physical differences existing between organisms and anorgana, consequently, do not lie in their material foundation; they do not arise from the different nature of the *elements* composing them, but from the different manner in which the latter are united by chemical *combination*. This different manner of combination gives rise to certain physical peculiarities, especially in density of substance, which at first sight seems to constitute a deep chasm between the two groups of bodies. Inorganic or inanimate natural bodies, such as crystals and the amorphous rocks, are in a state of density which we call the firm or solid state, and which we oppose to the liquid state of water and to the gaseous state of air. It is familiar to every one that these three different degrees of density, or states of aggregation of anorgana, are by no means peculiar to the different elements, but are the results of a certain degree of temperature. Every inorganic solid body, by increase of temperature, can be reduced to the liquid or melted state, and, by further heat, to the gaseous or elastic state. In the same way most gaseous bodies, by a proper decrease of temperature can first be converted into a liquid state, and further, into a solid state of density.

In opposition to these three states of density of anorgana, the living body of all organisms—*animals as well as plants*—is in an altogether peculiar fourth state of aggregation. It is neither solid like stone, nor liquid like water, but presents rather a medium between these two states, which may therefore be designated as the firm-fluid or swollen state of aggregation (viscid). In all living bodies, without exception, there is a certain quantity of water combined in a peculiar way with solid matter, and owing to this characteristic combination of water with solid matter we have that soft state of aggregation, neither solid nor liquid, which is of great importance in the mechanical explanation of the phenomena of life. Its cause lies essentially in the physical and chemical properties of a simple, indivisible, elementary substance, namely, *carbon* (Gen. Morph. i. 122–130).

Of all elements, carbon is to us by far the most important and interesting, because this simple substance plays the largest part in all animal and vegetable bodies known to us. It is that element which, by its peculiar tendency to form complicated combinations with the other elements, produces the greatest variety of chemical compounds, and among them the forms and

living substance of animal and vegetable bodies. Carbon is especially distinguished by the fact that it can unite with the other elements in infinitely
manifold relations of number and weight. By the combination of carbon with
three other elements, with oxygen, hydrogen, and nitrogen (to which generally sulphur, and frequently, also, phosphorus is added), there arise those
exceedingly important compounds which we have become acquainted with
as the first and most indispensable substratum of all vital phenomena, the
albuminous combinations, or albuminous bodies (protean matter).

We have before this (p. 185 [105]) become acquainted with the simplest
of all species of organisms in the Monera, whose entire bodies when completely developed consist of nothing but a semi-fluid albuminous lump; they
are organisms which are of the utmost importance for the theory of the first
origin of life. But most other organisms, also, at a certain period of their
existence—at least, in the first period of their life—in the shape of egg-cells
or germ-cells, are essentially nothing but simple little lumps of such albuminous formative matter, known as plasma, or protoplasma. They then
differ from the Monera only by the fact that in the interior of the albuminous corpuscle the cell-kernel, or nucleus, has separated itself from the surrounding cell-substance (protoplasma). As we have already pointed out, the
cells, with their simple attributes, are so many citizens, who by co-operation
and differentiation build up the body of even the most perfect organism;
this being, as it were, a cell republic (p. 301 [172]). The fully developed form
and the vital phenomena of such an organism are determined solely by the
activities of these small albuminous corpuscles.

It may be considered as one of the greatest triumphs of recent biology,
especially of the theory of tissues, that we are now able to trace the wonder
of the phenomena of life to these substances, and that we can demonstrate
the *infinitely manifold and complicated physical and chemical properties of
the albuminous bodies to be the real cause of organic or vital phenomena.* All
the different forms of organisms are simply and directly the result of the
combination of the different forms of cells. The infinitely manifold varieties
of form, size, and combination of the cells have arisen only gradually by the
division of labour, and by the gradual adaptation of the simple homogeneous
lumps of plasma, which originally were the only constituents of the cell-
mass. From this it follows of necessity that the fundamental phenomena of
life—nutrition and generation—in their highest manifestations, as well as
in their simplest expressions, must also be traced to the material nature of
that albuminous formative substance. The other vital activities are gradually evolved from these two. Thus, then, the general explanation of life is
now no more difficult to us than the explanation of the physical properties of
inorganic bodies. All vital phenomena and formative processes of organisms
are as directly dependent upon the chemical composition and the physical
forces of organic matter as the vital phenomena of inorganic crystals—that
is, the process of *their* growth and *their* specific formation—are the direct

results of their chemical composition and of their physical condition. The *ultimate causes*, it is true, remain in *both* cases concealed from us. When gold and copper crystallize in a cubical, bismuth and antimony in a hexagonal, iodine and sulphur in a rhombic form of crystal, the occurrence is in reality neither more nor less mysterious to us than is every elementary process of organic formation, every self-formation of the organic cell. In this respect we can no longer draw a fundamental distinction between organisms and anorgana, a distinction of which, formerly, naturalists were generally convinced.

Let us secondly examine the agreements and differences which are presented to us in the *formation* of organic and inorganic natural bodies (Gen. Morph. i. 130). Formerly the simple structure of the latter and the composite structure of the former were looked upon as the principal distinction. The body of all organisms was supposed to consist of dissimilar or heterogeneous parts, of instruments or organs which worked together for the purposes of life. On the other hand, the most perfect anorgana, that is to say, crystals, were supposed to consist entirely of continuous or homogeneous matter. This distinction appears very essential. But it loses all importance through the fact that in late years we have become acquainted with the exceedingly remarkable and important Monera.[15] (Compare above, p. 185 [105].) The whole body of these most simple of all organisms—a semi-fluid, formless, and simple lump of albumen—consists, in fact, of only a single chemical combination, and is as perfectly simple in its structure as any crystal, which consists of a single inorganic combination, for example, of a metallic salt or of a silicate of the earths and alkalies.

As naturalists believed in differences in the inner structure or composition, so they supposed themselves able to find complete differences in the external forms of organisms and anorgana, especially in the mathematically determinable crystalline forms of the latter. Certainly crystallization is preeminently a quality of the so-called anorgana. Crystals are limited by plane surfaces, which meet in straight lines and at certain measurable angles. Animal and vegetable forms, on the contrary, seem at first sight to admit of no such geometrical determination. They are for the most part limited by curved surfaces and crooked lines, which meet at variable angles. But in recent times we have become acquainted, among Radiolaria[23] and among many other Protista, with a large number of lower organisms, whose body, in the same way as crystals, may be traced to a mathematically determinable fundamental form, and whose form in its whole, as well as in its parts, is bounded by definite geometrically determinable planes and angles. In my general doctrine of *Fundamental Forms, or Promorphology*, I have given detailed proofs of this, and at the same time established a general system of forms, the ideal stereometrical type-forms, which explain the real forms of inorganic crystals, as well as of organic individuals (Gen. Morph. i. 375–574). Moreover, there are also perfectly amorphous organisms, like the Monera, Amoeba, etc., which change their forms every moment, and in which we are as little able to point out a definite fundamental form as in the case of the shapeless or amorphous anorgana, such as non-crystallized stones, deposits, etc. We are consequently unable to find any essential difference in the external forms or the inner structure of anorgana and organisms.

Thirdly, let us turn to the *forces* or the *phenomena of motion* of these two different groups of bodies (Gen. Morph. i. 140). Here we meet with the greatest difficulties. The vital phenomena, known as a rule only in the highly developed organisms, in the more perfect animals and plants, seem there so mysterious, so wonderful, so peculiar, that most persons are decidedly of opinion that in inorganic nature there occurs nothing at all similar, or in the least degree comparable to them. Organisms are for this very reason called animate, and the anorgana, inanimate natural bodies. Hence, even so late as the commencement of the present century, the science which investigates the phenomena of life, namely physiology, retained the erroneous idea that the physical and chemical properties of matter were not sufficient for explaining these phenomena. In our own day, especially during the last ten years, this idea may be regarded as having been completely refuted. In physiology, at least, it has now no place. It now never occurs to a physiologist to consider any of the vital phenomena as the result of a mysterious *vital force*, of an active power working for a definite purpose, standing outside of matter, and, so to speak, taking only the physico-chemical forces into its service. Modern physiology has arrived at the strictly monistic conviction that all of the vital phenomena, and, above all, the two fundamental phenomena of nutrition and propagation are purely physico-chemical processes, and directly dependent on the material nature of the organism, just as all the physical and chemical qualities of every crystal are determined solely by its material composition. Now, as the elementary substance which determines the peculiar material composition of organisms is carbon, we must ultimately reduce all vital phenomena, and, above all, the two fundamental phenomena of nutrition and propagation to the properties of the carbon. *The peculiar-chemico-physical properties, and especially the semi-fluid state of aggregation, and the easy decomposibility of the exceedingly composite albuminous combinations of carbon, are the mechanical causes of those peculiar phenomena of motion which distinguish organisms from anorgana, and which in a narrow sense are usually called "life."*

In order to understand this "*carbon theory*," which I have established in detail in the second book of my General Morphology, it is necessary, above all things, closely to examine those phenomena of motion which are common to both groups of natural bodies. First among them is the *process of growth*. If we cause any inorganic solution of salt slowly to evaporate, crystals are formed in it, which slowly increase in size during the continued evaporation of the water. This process of growth arises from the fact that new particles continually pass over from the fluid state of aggregation into the solid, and, according to certain laws, deposit themselves upon the firm kernel of the crystal already formed. From such an apposition of particles arise the mathematically definite crystalline shapes. In like manner the growth of organisms takes place by the accession of new particles. The only difference is that in the growth of organisms, in consequence of their semi-fluid state of aggregation, the newly-added particles penetrate into the interior of the organism (inter-susception), whereas anorgana receive homogeneous matter from without only by apposition or an addition of new particles to the surface. This important difference of growth by inter-susception and by apposition is obviously only the necessary and direct result of the different conditions of density or state of aggregation in organisms and anorgana.

Unfortunately I cannot here follow in detail the various exceedingly interesting parallels and analogies which occur between the formation of the most perfect anorgana, the crystals, and the formation of the simplest organisms, the Monera and their next kindred forms. For this I must refer to a minute comparison of organisms and anorgana, which I have carried out in the fifth chapter of my General Morphology (Gen. Morph. i. 111–160). I have there shown in detail that there exist no complete differences between organic and inorganic natural bodies, neither in respect to form and structure, nor in respect to matter and force; and that the actually existing differences are dependent upon the peculiar nature of the *carbon*; and that there exists no insurmountable chasm between organic and inorganic nature. We can perceive this most important fact very clearly if we examine and compare the origin of the forms in crystals and in the simplest organic individuals. In the formation of crystal individuals, two different counteracting formative tendencies come into operation. The *inner constructive force*, or the inner formative tendency, which corresponds to the Heredity of organisms, in the case of the crystal is the direct result of its material constitution or of its chemical composition. The form of the crystal, so far as it is determined by this inner original formative tendency, is the result of the specific and definite way in which the smallest particles of the crystallizing matter unite together in different directions according to law. That independent inner formative force, which is directly inherent in the matter itself, is directly counteracted by a second formative force. The *external constructive force*, or the external formative tendency, may be called Adaptation in crystals as well as in organisms. Every crystal individual during its formation, like every organic individual, must submit and adapt itself to the surrounding influences and conditions of existence of the outer world. In fact, the form and size of every crystal is dependent upon its whole surroundings, for example, upon the vessel in which the crystallization takes place, upon the temperature and the pressure of the air under which the crystal is formed, upon the presence or absence of heterogeneous bodies, etc. Consequently, the form of every single crystal, like the form of every single organism, is the result of the interaction of two opposing factors—the *inner* formative tendency, which is determined by the chemical constitution of the *matter itself*, and of the *external* formative tendency, which is dependent upon the influence of *surrounding* matter. Both these constructive forces interact similarly also in the organism, and, just as in the crystal, are of a purely mechanical nature and directly inherent in the substance of the body. If we designate the growth and the formation of organisms as a process of life, we may with equal reason apply the same term to the developing crystal. The teleological conception of nature, which looks upon organisms as machines of creation arranged for a definite purpose, must logically acknowledge the same also in regard to the forms of crystals. The differences which exist between the simplest organic individuals and inorganic crystals are determined by the *solid* state of aggregation of the latter, and by the *semi-fluid* state of the former. Beyond that the causes producing form are exactly the same in both. This conviction forces itself upon us most clearly, if we compare the exceedingly remarkable phenomena of growth, adaptation, and the "correlation of parts" of developing crystals with the corresponding phenomena of the origin of the simplest organic individuals (Monera and cells). The analogy between the two is so great that, in reality, no accurate boundary can be drawn. In my General Morphology I have quoted in support of this a number of striking facts (Gen. Morph. i. 146, 156, 158.)

If we vividly picture to ourselves this *"unity of organic and inorganic nature"* this essential agreement of organisms and anorgana in matter, form, and force, and if we bear in mind that we are not able to establish any one fundamental distinction between these two groups of bodies (as was formerly generally assumed), then the question of spontaneous generation will lose a great deal of the difficulty which at first seems to surround it. Then the development of the first organism out of inorganic matter will appear a much more easily conceivable and intelligible process than has hitherto been the case, whilst an artificial absolute barrier between organic or animate, and inorganic or inanimate nature was maintained.

In the question of *spontaneous generation, or archigony*, which we can now answer more definitely, it must be borne in mind that by this conception we understand generally the *non-parental generation of an organic individual*, the origin of an organism independent of a parental or producing organism. It is in this sense that on a former occasion (p. 183 [104]) I mentioned spontaneous generation (archigony) as opposed to parental generation or propagation (tocogony). In the latter case the organic individual arises by a greater or less portion of an already existing organism separating itself and growing independently. (Gen. Morph. ii. 32.)

In spontaneous generation, which is often also called original generation (generatio spontanea, aequivoca, primaria etc.), we must first distinguish two essentially different kinds, namely, *autogeny* and *plasmogeny*. By *autogeny* we understand the origin of a most simple organic individual in an *inorganic formative fluid*, that is, in a fluid which contains the fundamental substances for the composition of the organism dissolved in simple and loose combinations (for example, carbonic acid, ammonia, binary salts, etc.). On the other hand, we call spontaneous generation *plasmogeny* when the organism arises in an *organic formative fluid*, that is, in a fluid which contains those requisite fundamental substances dissolved in the form of complicated and fluid combinations of carbon (for example, albumen, fat, hydrate of carbon, etc.). (Gen. Morph. i. 174, ii. 33.)

Neither the process of autogeny, nor that of plasmogeny, has yet been directly observed with perfect certainty. In early, and also in more recent times, numerous and interesting experiments have been made as to the possibility or reality of spontaneous generation. Almost all these experiments refer not to autogeny, but to plasmogeny, to the origin of an organism out of already formed organic matter. It is evident, however, that this latter process is only of subordinate interest for our history of creation. It is much more important for us to solve the question, "Is there such a thing as autogeny? Is it possible that an organism can arise, not out of pre-existing organic, but out of purely inorganic, matter?" Hence we can quietly lay aside all the numerous experiments which refer only to plasmogeny, which have been carried on very zealously during the last ten years, and which for the most part have had a negative result. For even supposing that the reality of plasmogeny were strictly proved, still autogeny would not be explained by it.

The experiments on autogeny have likewise as yet furnished no certain and positive result. Yet we must at the outset most distinctly protest against the notion that these experiments have proved the impossibility of spontaneous generation in general. Most naturalists who have endeavoured to decide this question experimentally, and who, after having employed all possible precautionary measures, under well-ascertained conditions, have seen no organisms come into being, have straightway made the assertion, on the ground of these negative results: "That it is altogether impossible for organisms to come into existence by themselves without parental generation." This hasty and inconsiderate assertion they have supported by the negative results of their experiments, which, after all, could prove nothing except that, under these or those highly artificial circumstances created by the experimenters themselves, no organism was developed. From these experiments, which have been for the most part made under the most unnatural conditions, and in a highly artificial manner, we can by no means draw the conclusion that spontaneous generation in general is impossible. The impossibility of such a process can, in fact, never be proved. For how can we know that in remote primaeval times there did not exist conditions quite different from those at present obtaining, and which may have rendered spontaneous generation possible? Indeed, we can even positively and with full assurance maintain that the general conditions of life in primaeval times must have been entirely different from those of the present time. Think only of the fact that the enormous masses of carbon which we now find deposited in the primary coal mountains were first reduced to a solid form by the action of vegetable life, and are the compressed and condensed remains of innumerable vegetable substances, which have accumulated in the course of many millions of years. But at the time when, after the origin of water in a liquid state on the cooled crust of the earth, organisms were first formed by spontaneous generation, those immeasurable quantities of carbon existed in a totally different form, probably for the most part dispersed in the atmosphere in the shape of carbonic acid. The whole composition of the atmosphere was therefore extremely different from the present. Further, as may be inferred upon chemical, physical, and geological grounds, the density and the electrical conditions of the atmosphere were quite different. In like manner the chemical and physical nature of the primaeval ocean, which then continuously covered the whole surface of the earth as an uninterrupted watery sheet, was quite peculiar. The temperature, the density, the amount of salt, etc., must have been very different from those of the present ocean. In any case, therefore, even if we do not know anything more about it, there remains to us the supposition, which can at least not be disputed, that at that time, under conditions quite different from those of to-day, a spontaneous generation, which now is perhaps no longer possible, may have taken place.

But it is necessary to add here that, by the recent progress of chemistry and physiology, the mysterious and miraculous character which at first seems to belong to this much disputed and yet inevitable process of spontaneous generation, has been to a great extent, or almost entirely, destroyed. Not fifty years ago, all chemists maintained that we were unable to produce artificially in our laboratories any complicated combination of carbon, or so-called "organic combination." The mystic "vital force" alone was supposed to be able to produce these combinations. When, therefore, in 1828, Wöhler, in Göttingen, for the first time refuted this dogma, and exhibited pure "organic" urea, obtained in an artificial manner from a purely inorganic body (cyanate of ammonium), it caused the greatest surprise and astonishment. In more recent times, by the progress of synthetic chemistry, we have succeeded in producing in our laboratories a great variety of similar "organic" combinations of carbon, by purely artificial means—for example alcohol, acetic acid, formic acid. Indeed, many exceedingly complicated combinations of carbon are now artificially produced, so that there is every likelihood, sooner or later, of our producing artificially the most complicated, and at the same time the most important of all, namely, the albuminous combinations, or plasma-bodies. By the consideration of this probability, the deep chasm which was formerly and generally believed to exist between organic and inorganic bodies is almost or entirely removed, and the way is paved for the conception of spontaneous generation.

Of still greater, nay, the very greatest importance to the hypothesis of spontaneous generation are, finally, the exceedingly remarkable *Monera*, those creatures which we have already so frequently mentioned, and which are not only the simplest of all observed organisms, but even the simplest of all imaginable organisms. I have already described these wonderful *"organisms without organs,"* when examining the simplest phenomena of propagation and inheritance. We already know seven different genera of these Monera, some of which live in fresh water, others in the sea (compare above, p. 184 [194]; also Plate I. and its explanation in the Appendix). In a perfectly developed and freely motile state, they one and all present us with nothing but a simple little lump of an albuminous combination of carbon. The individual genera and species differ only a little in the manner of propagation and development, and in the way of taking nourishment. Through the discovery of these organisms, which are of the utmost importance, the supposition of a spontaneous generation loses most of its difficulties. For as all trace of organization—all distinction of heterogeneous parts—is still wanting in them, and as all the vital phenomena are performed by one and the same homogeneous and formless matter, we can easily imagine their origin by spontaneous generation. If this happens through *plasmogeny*, and if plasma capable of life already exists, it then only needs to individualize itself in the same way as the mother liquor of crystals individualizes itself in crystallization. If, on the other hand, the spontaneous generation of the Monera takes place by true *autogeny*, then it is further requisite that that plasma capable of life, that primaeval mucus, should be formed out of simpler combinations of carbon. As we are now able artificially to produce, in our laboratories, combinations of carbon similar to this in the complexity of their constitution, there is absolutely no reason for supposing that there are not conditions in free nature also, in which such combinations could take place. Formerly, when the doctrine of spontaneous generation was advocated, it failed at once to obtain adherents on account of the composite structure of the simplest organisms then known. It is only since we have discovered the exceedingly important Monera, only since we have become acquainted in them with organisms not in any way built up of distinct organs, but which consist solely of a single chemical combination, and yet grow, nourish, and propagate themselves, that this great difficulty has been removed, and the hypothesis of spontaneous generation has gained a degree of probability which entitles it to fill up the gap existing between Kant's cosmogony and Lamarck's Theory of Descent. Even among the Monera at present known there is a species which probably, even now, always comes into existence by spontaneous generation. This is the wonderful *Bathybius haeckelii*, discovered and described by Huxley. As I have already mentioned (p. 184 [105]), this Moneron is found in the greatest depths of the sea, at a depth of between 12,000 and 24,000 feet, where it covers the ground partly as retiform threads and plaits of plasma, partly in the form of larger or smaller irregular lumps of the same material.*

---

* We must wait for fuller information on the subject of *Bathybius*, at the hands of the naturalists of the *Challenger* expedition, before accepting it finally as a distinct organism.— E.R.L. (translator).

Only such homogeneous organisms as are yet not differentiated, and are similar to inorganic crystals in being homogeneously composed of one single substance, could arise by spontaneous generation, and could become the primaeval parents of all other organisms. In their further development we have pointed out that the most important process is the formation of a *kernel* or *nucleus* in the simple little lump of albumen. We can conceive this to take place in a purely physical manner, by the condensation of the innermost central part of the albumen. The more solid central mass, which at first gradually shaded off into the peripheral plasma, becomes sharply separated from it, and thus forms an independent, round, albuminous corpuscle, the kernel; and by this process the Moneron becomes a *cell*. Now, it must have become evident from our previous chapters, that the further development of all other organisms out of such a cell presents no difficulty, for every animal and every plant, in the beginning of its individual life, is a simple cell. Man, as well as every other animal, is at first nothing but a simple egg-cell, a single lump of mucus, containing a kernel (p. 297 [169], Fig. 5).

In the same way as the kernel of the organic cell arose in the interior or central mass of the originally homogeneous lump of plasma, by separation, so, too, the first *cell-membrane* was formed on its surface. This simple, but most important process, as has already been remarked, can likewise be explained in a purely physical manner, either as a chemical deposit, or as a physical condensation in the uppermost stratum of the mass, or as a secretion. One of the first processes of adaptation effected by the Moneron originating by spontaneous generation must have been the condensation of an external crust, which as a protecting covering shut in the softer interior from the hostile influences of the outer world. As soon as, by condensation of the homogeneous Moneron, a cell-kernel arose in the interior and a membrane arose on the surface, all the fundamental parts of the unit were furnished, out of which, by infinitely manifold repetition and combination, as attested by actual observation, the body of higher organisms is constructed.

As has already been mentioned, our whole understanding of an organism rests upon the cell theory established thirty years ago by Schleiden and Schwann. According to it, every organism is either a simple cell or a cell-community, a republic of closely connected cells. All the forms and vital phenomena of every organism are the collective result of the forms and vital phenomena of all the single cells of which it is composed. By the recent progress of the cell theory it has become necessary to give the elementary organisms, that is, the "organic" individuals of the first order, which are usually designated as *cells*, the more general and more suitable name of *form-units*, or *plastids*. Among these form-units we distinguish two main groups, namely, the cytods and the genuine cells. The *cytods* are, like the Monera, pieces of plasma without a kernel (p. 186 [107], Fig. 1). *Cells*, on the other hand, are pieces of plasma containing a kernel or nucleus (p. 188 [108], Fig. 2). Each of these two main groups of plastids is again divided into two subordinate groups, according as they possess or do not possess an external covering (skin, shell, or membrane). We may accordingly distinguish the following four grades or species of plastids, namely: 1. *Simple cytods* (p. 186 [107], Fig. 1 *A*); 2. *Encased cytods*; 3. *Simple cells* (p. 188 [108], Fig. 2 *B*); 4. *Encased cells* (p. 188 [108], Fig. 2 *A*). (Gen. Morph. i. 269–289.)

Concerning the relation of these four forms of plastids to spontaneous generation, the following is the most probable:—1. The *simple cytods* (Gymnocytoda), naked particles of plasma without kernel, like the still living Monera, are the only plastids which directly come into existence by spontaneous generation. 2. The *enclosed cytods* (Lepocytoda), particles of plasma without kernel, which are surrounded by a covering (membrane or shell), arose out of the simple cytods either by the condensation of the outer layers of plasma or by the secretion of a covering. 3. The *simple cells* (Gymnocyta), or naked cells, particles of plasma with kernel, but without covering, arose out of the simple cytods by the condensation of the innermost particles of plasma into a kernel, or nucleus, by differentiation of a central kernel and peripheral cell-substance. 4. The *enclosed cells* (Lepocyta), or testaceous cells, particles of plasma with kernel and an outer covering (membrane or shell), arose either out of the enclosed cytods by the formation of a kernel, or out of the simple cells by the formation of a membrane. All the other forms of form-units, or plastids, met with, besides these, have only subsequently arisen out of these four fundamental forms by natural selection, by descent with adaptation, by differentiation and transformation.

By this *theory of plastids*, by deducing all the different forms of plastids, and hence, also, all organisms composed of them, from the Monera, we obtain a simple and natural connection in the whole series of the development of nature. The origin of the first Monera by spontaneous generation appears to us as a simple and necessary event in the process of the development of the earth. We admit that this process, as long as it is not directly observed or repeated by experiment, remains a pure hypothesis. But I must again say that this hypothesis is indispensable for the consistent completion of the non-miraculous history of creation, that it has absolutely nothing forced or miraculous about it, and that certainly it can never be positively refuted. It must be taken into consideration that the process of spontaneous generation, even if it still took place daily and hourly, would in any case be exceedingly difficult to observe and establish with absolute certainty as such. With regard to the Monera, we find ourselves placed before the following alternative: *either* they are actually directly derived from pre-existing, or "created," most ancient Monera, and in this case they would have had to propagate themselves unchanged for many millions of years, and to have maintained their original form of simple particles of plasma; *or*, the *present* Monera have originated much later in the course of the organic history of the earth, by repeated acts of spontaneous generation, and in this case spontaneous generation may take place now as well as then. The latter supposition has evidently much more probability on its side than the former.

If we do not accept the hypothesis of spontaneous generation, then at this one point of the history of development we must have recourse to the miracle of a *supernatural creation*. The Creator must have created the first organism, or a few first organisms, from which all others are derived, and as such he must have created the simplest Monera, or primaeval cytods, and given them the capability of developing further in a mechanical way. I leave it to each one of my readers to choose between this idea and the hypothesis of spontaneous generation. To me the idea that the Creator should have in this one point arbitrarily interfered with the regular process of development of matter, which in all other cases proceeds entirely without his interposition, seems to be just as unsatisfactory to a believing mind as to a scientific intellect. If, on the other hand, we assume the hypothesis of spontaneous generation for the origin of the first organisms, which in consequence of reasons mentioned above, and especially in consequence of the discovery of the Monera, has lost its former difficulty, then we arrive at the establishment of an uninterrupted natural connection between the development of the earth and the organisms produced on it, and, in this last remaining lurking-place of obscurity, we can proclaim the *unity of all Nature, and the unity of her laws of Development* (Gen. Morph. i. 164).

CHAPTER XIV.

MIGRATION AND DISTRIBUTION OF ORGANISMS. CHOROLOGY AND THE ICE-PERIOD OF THE EARTH.

Chorological Facts and Causes.—Origin of most Species in one Single Locality: "Centres of Creation."—Distribution by Migration.—Active and Passive Migrations of Animals and Plants.—Means of Transport.—Transport of Germs by Water and by Wind.—Continual Change of the Area of Distribution by Elevations and Depressions of the Ground.—Chorological Importance of Geological Processes.—Influence of the Change of Climate.—Ice or Glacial Period.—Its Importance to Chorology.—Importance of Migrations for the Origin of New Species.—Isolation of Colonists.—Wagner's Law of Migration.—Connection between the Theory of Migration and the Theory of Selection.—Agreement of its Results with the Theory of Descent.

As I have repeatedly said, but cannot too much emphasize, the actual value and invincible strength of the Theory of Descent does not lie in its explaining this or that single phenomenon, but in the fact that it explains *all* biological phenomena, that it makes *all* botanical and zoological series of phenomena intelligible in their relations to one another. Hence every thoughtful investigator is the more firmly and deeply convinced of its truth the more he advances from single biological observations to a general view of the whole domain of animal and vegetable life. Let us now, starting from this comprehensive point of view, survey a biological domain, the varied and complicated phenomena of which may be explained with remarkable simplicity and clearness by the theory of selection. I mean *Chorology*, or the theory of the *local distribution of organisms over the surface of the earth*. By this I do not only mean the *geographical* distribution of animal and vegetable species over the different parts and provinces of the earth, over continents and islands, seas, and rivers; but also their *topographical* distribution in a *vertical* direction, their ascending to the heights of mountains, and their descending into the depths of the ocean. (Gen. Morph. ii. 286.)

The strange chorological series of phenomena which show the horizontal distribution of organisms over parts of the earth, and their vertical distribution in heights and depths, have long since excited general interest. In recent times Alexander Humboldt[39] and Frederick Schouw have especially discussed the geography of plants, and Berghaus and

Schmarda the geography of animals, on a large scale. But although these and several other naturalists have in many ways increased our knowledge of the distribution of animal and vegetable forms, and laid open to us a new domain of science, full of wonderful and interesting phenomena, yet Chorology as a whole remained, as far as their labours were concerned, only a desultory knowledge of a mass of individual *facts*. It could not be called a science as long as the *causes* for the explanation of these facts were wanting. These causes were first disclosed by the theory of selection and its doctrine of the *migrations* of animal and vegetable species, and it is only since the works of Darwin and Wallace that we have been able to speak of an independent *science of Chorology*.

If all the phenomena of the geographical and topographical distribution of organisms are examined by themselves, without considering the gradual development of species, and if at the same time, following the customary superstition, the individual species of animals and plants are considered as forms independently created and independent of one another, then there remains nothing for us to do but to gaze at those phenomena as a confused collection of incomprehensible and inexplicable miracles. But as soon as we leave this low stand-point, and rise to the height of the theory of development, by means of the supposition of a blood-relationship between the different species, then all at once a clear light falls upon this strange series of miracles, and we see that all chorological facts can be understood quite simply and clearly by the supposition of a common descent of the species, and their passive and active migrations.

The most important principle from which we must start in chorology, and of the truth of which we are convinced by due examination of the theory of selection, is that, as a rule, every animal and vegetable species has arisen only *once* in the course of time and only in *one* place on the earth—its so-called "centre of creation"—by natural selection. I share this opinion of Darwin's unconditionally, in respect to the great majority of higher and perfect organisms, and in respect to most animals and plants in which the division of labour, or differentiation of the cells and organs of which they are composed, has attained a certain stage. For it is quite incredible, or could at best only be an exceedingly rare accident, that all the manifold and complicated circumstances—all the different conditions of the struggle for life, which influence the origin of a new species by natural selection—should have worked together in exactly the same agreement and combination more than once in the earth's history, or should have been active at the same time at several different points of the earth's surface.

On the other hand, I consider it to be very probable that certain exceedingly imperfect organisms of the simplest structure, forms of species of an exceedingly indifferent nature, as, for example, many single-celled Protista, but especially the Monera, the simplest of them all, should have several times or simultaneously arisen in their specific form in several parts of the earth. For the few and very simple conditions by which their specific form was changed in the struggle for life may surely have often been repeated, in the course of time, independently in different parts of the earth. Further, those higher specific forms also, which have not arisen by natural selection, but by *hybridism* (the previously-mentioned hybrid species, pp. 147 [83] and 275 [157]), may have repeatedly arisen anew in different localities. As, however, this proportionately small number of organisms does not especially interest us here, we may, in respect of chorology, leave them alone, and need only take into consideration the distribution of the great majority of animal and vegetable species in regard to which the *single origin of every species in a single locality*, in its so-called "central point of creation," can be considered as tolerably certain.

Every animal and vegetable species from the beginning of its existence has possessed the tendency to spread beyond the limited locality of its origin, beyond the boundary of its "centre of creation," or, in other words, beyond its *primaeval home*, or its natal place. This is a necessary consequence of the relations of population and over-population (pp. 161 [91] and 256 [146]). The more an animal or vegetable species increases, the less is its limited natal place sufficient for its sustenance, and the fiercer the struggle for life; the more rapid the *over-population* of the natal spot, the more it leads to *emigration*. These *migrations* are common to all organisms, and are the real cause of the wide distribution of the different species of organisms over the earth's surface. Just as men leave over-crowded states, so all animals and plants migrate from their over-crowded primaeval homes.

Many distinguished naturalists, especially Lyell[11] and Schleiden, have before this repeatedly drawn attention to the great importance of these very interesting migrations of organisms. The means of transport by which they are effected are extremely varied. Darwin has discussed these most excellently in the eleventh and twelfth chapters of his work, which are exclusively devoted to "geographical distribution." The means of transport are partly active, partly passive; that is to say, the organism effects its migration partly by free locomotion due to its own activity, and partly by the movements of other natural bodies in which it has no active share.

It is self-evident that *active migrations* play the chief part in animals able to move freely. The more freely an animal's organization permits it to all move in directions, the more easily the animal species can migrate, and the more rapidly it will spread over the earth. *Flying* animals are of course most favoured in this respect, among vertebrate animals especially birds, and among articulated animals, insects. These two classes, as soon as they came into existence, can have more easily spread over the whole earth than any other animal, and this fact partly explains the extraordinary uniformity of structure which characterizes these two great classes of animals. For, although they contain an exceedingly large number of different species, and although the insect class alone is said to possess more different species than all other classes of animals together, yet all the innumerable species of insects, and in like manner, also, the different species of birds, agree most strikingly in all essential peculiarities of their organization. Hence, in the class of insects, as well as in that of birds, we can distinguish only a very small number of large natural groups or orders, and these few orders differ but very little from one another in their internal structure. The orders of birds with their numerous species are not nearly as distinct from one another as the orders of the mammalian class, containing much fewer species; and the orders of insects, which are extremely rich in genera and species, resemble one another much more closely in their internal structure than do the much smaller orders of the crab class. The general parallelism between birds and insects is also very interesting in relation to systematic zoology; and the great importance of their richness in forms, for scientific morphology, lies in the fact that they show us how, within the narrowest anatomical sphere, and without profound changes of the essential internal organization, the greatest variety in external bodily forms can be attained. The reason of this is evidently their flying mode of life and their free locomotion. In consequence of this birds, as well as insects, have spread very rapidly over the whole surface of the earth, have settled in all possible localities inaccessible to other animals, and variously modified their specific form by superficial adaptation to particular local relations.

Next to the flying animals, those animals, of course, have spread most quickly and furthest which were next best able to migrate, that is, the best runners among the inhabitants of the land, and the best swimmers among the inhabitants of the water. However, the power of such active migrations is not confined to those animals which throughout life enjoy free locomotion. For the fixed animals also, such as corals, tubicolous worms, sea-squirts, lily encrinites, sea-acorns, barnacles, and many other lower animals which adhere to seaweeds, stones, etc., enjoy, at least at an early period of life, free locomotion. They all migrate before they adhere to anything. Their first free locomotive condition of early life is generally that of a "ciliated" larva, a roundish, cellular corpuscle, which, by means of a garb of movable "flimmer-hairs," (Latin, "cilia") swarms about in the water and bears the name of Planula.

But the power of free locomotion, and hence, also, of active migration, is not confined to animals alone, but many plants likewise enjoy it. Many lower aquatic plants, especially the class of the Tangles (Algae), swim about freely in the water in early life, like the lower animals just mentioned, by means of a vibratile hairy coat, a vibrating whip, or a covering of tremulous fringes, and only at a later period adhere to objects. Even in the case of many higher plants, which we designate as creepers and climbing plants, we may speak of active migration. Their elongated stalks and perennial roots creep or climb during their long process of growth to new positions, and by means of their widespread branches they acquire new habitations, to which they attach themselves by buds, and bring forth new colonies of individuals of their species.

Influential as these active migrations of most animals and many plants are, yet alone they would by no means be sufficient to explain the chorology of organisms. *Passive migrations* have ever been by far the more important, and of far greater influence, in the case of most plants and in that of many animals. Such passive changes of locality are produced by extremely numerous causes. Air and water in their eternal motion, wind and waves with their manifold currents, play the chief part. The wind in all places and at all times raises light organisms, small animals and plants, but especially their young germs, animal eggs and plant seeds, and carries them far over land and seas. Where they fall into the water they are seized by currents or waves and carried to other places. It is well known, from numerous examples, how far in many cases trunks of trees, hard shelled fruits, and other not readily perishable portions of plants are carried away from their original home by the course of rivers and by the currents of the sea. Trunks of palm trees from the West Indies are brought by the Gulf Stream to the British and Norwegian coasts. All large rivers bring down driftwood from the mountains, and frequently alpine plants are carried from their home at the source of the river into the plains, and even further, down to the sea. Frequently numerous inhabitants live between the roots of the plants thus carried down, and between the branches of the trees thus washed away there are various inhabitants which have to take part in the passive migration. The bark of the tree is covered with mosses, lichens, and parasitic insects. Other insects, spiders, etc., even small reptiles and mammals, are hidden within the hollow trunk or cling to the branches. In the earth adhering to the fibres of the roots, in the dust lying in the cracks of the bark, there are innumerable germs of smaller animals and plants. Now, if the trunk thus washed away lands safely on a foreign shore or on a distant island, the guests who had to take part in the involuntary voyage can leave their boat and settle in the new country. A very remarkable kind of water-transport is formed by the floating icebergs which annually become loosened from the eternal ice of the Polar Sea. Although these cold regions are thinly peopled, yet many of their inhabitants, who were accidentally upon an iceberg while it was becoming loosened, are carried away with it by the currents, and landed on warmer shores. In this manner, by means of loosened blocks of ice from the northern Polar Sea, often whole populations of small animals and plants have been carried to the northern shores of Europe and America. Nay, even polar foxes and polar bears have been carried in this way to Iceland and to the British Isles.

Transport by air is no less important than transport by water in this matter of passive migration. The dust covering our streets and roofs, the earth lying on dry fields and dried-up pools, the light moist soil of forests, in short, the whole surface of the globe contains millions of small organisms and their germs. Many of these small animals and plants can without injury become completely dried up, and awake again to life as soon as they are moistened. Every gust of wind raises up with the dust innumerable little creatures of this kind, and often carries them away to other places miles off. But even larger organisms, and especially their germs, may often make distant passive journeys through the air. The seeds of many plants are provided with light feathery processes, which act as parachutes and facilitate their flight in the air, and prevent their falling. Spiders make journeys of many miles through the air on their fine filaments, their so-called gossamer threads. Young frogs are frequently raised by whirlwinds into the air by thousands, and fall down in a distant part as a "shower of frogs." Storms may carry birds and insects across half the earth's circumference. They drop in the United States, having risen in England. Starting from California, they only come to rest in China. But, again, many other organisms may make the journey from one continent to another together with the birds and insects. Of course all parasites, the number of which is legion, fleas, lice, mites, moulds, etc., migrate with the organisms upon which they live. In the earth which often remains sticking to the claws of birds there are also small animals and plants or their germs. Thus the voluntary or involuntary migration of a single larger organism may carry a whole small flora and fauna from one part of the earth to another.

Besides the means of transport here mentioned, there are many others which explain the distribution of animal and vegetable species over the large tracts of the earth's surface, and especially the general distribution of the so-called cosmopolitan species. But these alone would not nearly be sufficient to explain all chorological facts. How is it, for example, that many inhabitants of fresh water live in various rivers or lakes far away and quite apart from one another? How is it that many inhabitants of mountains, which cannot exist in plains, are found upon entirely separated and far distant chains of mountains? It is difficult to believe, and in many cases quite inconceivable, that these inhabitants of fresh water should have in any way, actively or passively, migrated over the land lying between the lakes, or that the inhabitants of mountains in any way, actively or passively, crossed the plains lying between their mountain homes. But here geology comes to our help, as a mighty ally, and completely solves these difficult problems for us.

The history of the earth's development shows us that the distribution of land and water on its surface is ever and continually changing. In consequence of geological changes of the earth's crust, *elevations* and *depressions* of the ground take place everywhere, sometimes more strongly marked in one place, sometimes in another. Even if they happen so slowly that in the course of centuries the seashore rises or sinks only a few inches, or even only a few lines, still they nevertheless effect great results in the course of long periods of time. And long—immeasurably long—periods of time have not been wanting in the earth's history. During the course of many millions of years, ever since organic life existed on the earth, land and water have perpetually struggled for supremacy. Continents and islands have sunk into the sea, and new ones have arisen out of its bosom. Lakes and seas have slowly been raised and dried up, and new water basins have arisen by the sinking of the ground. Peninsulas have become islands by the narrow neck of land which connected them with the mainland sinking into the water. The islands of an archipelago have become the peaks of a continuous chain of mountains by the whole floor of their sea being considerably raised.

Thus the Mediterranean at one time was an inland sea, when, in the place of the Straits of Gibraltar, an isthmus connected Africa with Spain. England, even during the more recent history of the earth, when man already existed, has repeatedly been connected with the European continent and been repeatedly separated from it. Nay, even Europe and North America have been directly connected. The South Sea at one time formed a large Pacific Continent, and the numerous little islands which now lie scattered in it were simply the highest peaks of the mountains covering that continent. The Indian Ocean formed a continent which extended from the Sunda Islands along the southern coast of Asia to the east coast of Africa. This large continent of former times Sclater, an Englishman, has called *Lemuria*, from the monkey-like animals which inhabited it, and it is at the same time of great importance from being the probable cradle of the human race, which in all likelihood here first developed out of anthropoid apes. The important proof which Alfred Wallace has furnished,[36] by the help of chorological facts, that the present Malayan Archipelago consists in reality of two completely different divisions, is particularly interesting. The western division, the Indo-Malayan Archipelago, comprising the large islands of Borneo, Java, and Sumatra, was formerly connected by Malacca with the Asiatic continent, and probably also with the Lemurian continent just mentioned. The eastern division, on the other hand, the Austro-Malayan Archipelago, comprising Celebes, the Moluccas, New Guinea, Solomon's Islands, etc., was formerly directly connected with Australia. Both divisions were formerly two continents separated by a strait, but they have now for the most part sunk below the level of the sea. Wallace, solely on the ground of his accurate chorological observations, has been able in the most acute manner to determine the position of this former strait, the south end of which passes between Balij and Lombok.

Thus, ever since liquid water existed on the earth, the boundaries of water and land have eternally changed, and we may assert that the outlines of continents and islands have never remained for an hour, nay, even for a minute, exactly the same. For the waves eternally and perpetually break on the edge of the coast, and whatever the land in these places loses in extent, it gains in other places by the accumulation of mud, which condenses into solid stone and again rises above the level of the sea as new land. Nothing can be more erroneous than the idea of a firm and unchangeable outline of our continents, such as is impressed upon us in early youth by defective lessons on geography, which are devoid of a geological basis.

I need hardly draw attention to the fact that these geological changes of the earth's surface have ever been exceedingly important to the migrations of organisms, and consequently to their Chorology. From them we learn to understand how it is that the same or nearly related species of animals and plants can occur on different islands, although they could not have passed through the water separating them, and how other species living in fresh water can inhabit different enclosed water-basins, although they could not have crossed the land lying between them. These islands were formerly mountain peaks of a connected continent, and these lakes were once directly connected with one another. The former were separated by geological depressions, the latter by elevations. Now, if we further consider how often and how unequally these alternating elevations and depressions occur on the different parts of the earth, and how, in consequence of this, the boundaries of the geographical tracts of distribution of species become changed, and if we further consider in what exceedingly various ways the active and passive migrations of organisms must have been influenced by them, then we shall be in a position to completely understand the great variety of the picture which is at present offered to us by the distribution of animal and vegetable species.

There is yet another important circumstance to be mentioned here, which is likewise of great importance for a complete explanation of this varied geographical picture, and which throws light upon many very obscure facts, which, without its help, we should not be able to comprehend. I mean the gradual *change of climate* which has taken place during the long course of the organic history of the earth. As we saw in our last chapter, at the beginning of organic life on the earth a much higher and more equal temperature must have generally prevailed than at present. The differences of zones, which in our time are so very striking, did not exist at all in those times. It is probable that for many millions of years but one climate prevailed over the whole earth, which very closely resembled, or even surpassed, the hottest tropical climate of the present day. The highest north which man has yet reached was then covered with palms and other tropical plants, the fossil remains of which are still found there. The temperature of this climate at a later period gradually decreased; but still the poles remained so warm that the whole surface of the earth could be inhabited by organisms. It was only at a comparatively very recent period of the earth's history, namely, at the beginning of the tertiary period, that there occurred, as it seems, the first perceptible cooling of the earth's crust at the poles, and through this the first differentiation or separation of the different zones of temperature or climatic zones. But the slow and gradual decrease of temperature continued to extend more and more within the tertiary period, until at last, at both poles of the earth, the first permanent ice caps were formed.

I need scarcely point out in detail how very much this change of climate must have affected the geographical distribution of organisms, and the origin of numerous new species. The animal and vegetable species, which, down to the tertiary period, had found an agreeable tropical climate all over the earth, even as far as the poles, were now forced either to adapt themselves to the intruding cold, or to flee from it. Those species which adapted and accustomed themselves to the decreasing temperature became new species simply by this very acclimatization, under the influence of natural selection. The other species, which fled from the cold, had to emigrate and seek a milder climate in lower latitudes. The tracts of distribution which had hitherto existed must by this have been vastly changed.

However, during the last great period of the earth's history, during the quaternary period (or diluvial period) succeeding the tertiary one, the decrease of the heat of the earth from the poles did not by any means remain stationary. The temperature fell lower and lower, nay, even far below the present degree. Northern and Central Asia, Europe, and North America from the north pole, were covered to a great extent by a connected sheet of ice, which in our part of the earth seems to have reached the Alps. In a similar manner the cold also advancing from the south pole covered a large portion of the southern hemisphere, which is now free from it, with a rigid sheet of ice. Thus, between these vast lifeless ice continents there remained only a narrow zone to which the life of the organic world had to withdraw. This period, during which man, or at least the human ape, already existed, and which forms the first period of the so-called *diluvial epoch*, is now universally known as the *ice* or *glacial period*.

The ingenious Carl Schimper is the first naturalist who clearly conceived the idea of the ice period, and proved the great extent of the former glaciation of Central Europe by the help of the so-called boulders, or erratic blocks of stone, as also by the "glacier tables." Louis Agassiz, stimulated by him, and considerably supported by the independent investigations of the eminent geologist Charpentier, afterwards undertook the task of carrying out the theory of the ice period. In England, the geologist Forbes distinguished himself in this matter, and also was the first to apply it to the theory of migrations and the geographical distribution of species dependent upon migration. Agassiz, however, afterwards injured the theory by his one-sided exaggeration, inasmuch as, from his partiality to Cuvier's theory of cataclysms, he endeavoured to attribute the destruction of the whole animate creation then existing, to the sudden coming on of the cold of the ice period and the "revolution" connected with it.

It is unnecessary here to enter into detail as to the ice period itself, and into investigations about its limits, and I may omit this all the more reasonably since the whole of our recent geological literature is full of it. It will be found discussed in detail in the works of Cotta,[31] Lyell,[30] Vogt,[27] Zittel,[32] etc. Its great importance to us here is that it helps us to explain the most difficult chorological problems, as Darwin has correctly perceived.

For there can be no doubt that this glaciation of the present temperate zones must have exercised an exceedingly important influence on the geographical and topographical distribution of organisms, and that it must have entirely changed it. While the cold slowly advanced from the poles towards the equator, and covered land and sea with a connected sheet of ice, it must of course have driven the whole living world before it. Animals and plants had to migrate if they wished to escape being frozen. But as at that time the temperate and tropical zones were probably no less densely peopled with animals and plants than at present, there must have arisen a fearful struggle for life between the latter and the intruders coming from the poles. During this struggle, which certainly lasted many thousands of years, many species must have perished and many become modified and been transformed into new species. The hitherto existing tracts of distribution of species must have become completely changed, and the struggle have been continued, nay, indeed, must have broken out anew and been carried on in new forms, when the ice period had reached and gone beyond its furthest point, and when in the post-glacial period the temperature again increased, and organisms began to migrate back again towards the poles.

In any case this great change of climate, whether a greater or less importance be ascribed to it, is one of those occurrences in the history of the earth which have most powerfully influenced the distribution of organic forms. But more especially one important and difficult chorological circumstance is explained by it in the simplest manner, namely, the specific agreement of many of our Alpine inhabitants with some of those living in polar regions. There is a great number of remarkable animal and vegetable forms which are common to these two far distant parts of the earth, and which are found nowhere in the wide plains lying between them. Their migration from the polar lands to the Alpine heights, or *vice versa*, would be inconceivable under the present climatic circumstances, or could be assumed at least only in a few rare instances. But such a migration could take place, nay, was obliged to take place, during the gradual advance and retreat of the ice-sheet. As the glaciation encroached from Northern Europe towards our Alpine chains, the polar inhabitants retreating before it—gentian, saxifrage, polar foxes, and polar hares—must have peopled Germany, in fact all Central Europe. When the temperature again increased, only a portion of these Arctic inhabitants returned with the retreating ice to the Arctic zones. Another portion of them climbed up the mountains of the Alpine chain instead, and there found the cold climate suited to them. The problem is thus solved in a most simple manner.

We have hitherto principally considered the *theory of the migrations* of organisms in so far as it explains the radiation of every animal and vegetable species from a single primaeval home, from a "central point of creation," and the dispersion of these species over a greater or less portion of the earth's surface. But these migrations are also of great importance to the theory of development, because we can perceive in them a very important means for the *origin of new species*. When animals and plants migrate they meet in their new home, in the same way as do human emigrants, with conditions which are more or less different from those which they have inherited throughout generations, and to which they have been accustomed. The emigrants must either submit and adapt themselves to these new conditions of life or they perish. By adaptation their peculiar specific character becomes the more changed the greater the difference between the new and the old home. The new climate, the new food, but above all, new neighbours in the forms of other animals and plants, influence and tend to modify the inherited character of the immigrant species, and if it is not hardy enough to resist the influences, then sooner or later a new species must arise out of it. In most cases this transformation of an immigrant species takes place so quickly under the influence of the altered struggle for life, that even after a few generations a new species arises from it.

Migration has an especial influence in this way on all organisms with separate sexes. For in them the origin of new species by natural selection is always rendered difficult, or delayed, by the fact that the modified descendants occasionally again mix sexually with the unchanged original form, and thus by crossing return to the first form. But if such varieties have migrated, if great distances or barriers to migration—seas, mountains, etc.—have separated them from the old home, then the danger of a mingling with the primary form is prevented, and the isolation of the emigrant form, which becomes a new species by adaptation, prevents its breeding with the old stock, and hence prevents its return in this way to the original form.

The importance of migration for the isolation of newly-originating species and the prevention of a speedy return to the primary form has been especially pointed out by the philosophic traveller, Moritz Wagner, of Munich. In a special treatise on "Darwin's Theory and the Law of the Migration of Organisms,"[40] Wagner gives from his own rich experience a great number of striking examples which confirm the theory of migration set forth by Darwin in the eleventh and twelfth chapters of his book, where he especially discusses the effect of the complete isolation of emigrant organisms in the origin of new species. Wagner sets forth the simple causes which have "locally bounded the form and founded its typical difference," in the following three propositions:—1. The greater the total amount of change in the hitherto existing conditions of life which the emigrating individuals find on entering a new territory, the more intensely must the innate variability of every organism manifest itself. 2. The less this increased individual variability of organisms is disturbed in the peaceful process of reproduction by the mingling of numerous subsequent immigrants of the same species, the more frequently will nature succeed, by intensification and transmission of the new characteristics, in forming a new variety or race, that is, a commencing species. 3. The more advantageous the changes experienced by the individual organs are to the variety, the more readily will it be able to adapt itself to the surrounding conditions; and the longer the undisturbed breeding of a commencing variety of colonists in a new territory continues without its mingling with subsequent immigrants of the same species, the oftener a new species will arise out of the variety.

Every one will agree with these three propositions of Moritz Wagner's. But we must consider his view, that the migration and the subsequent isolation of the emigrant individuals is a *necessary* condition for the origin of new species, to be completely erroneous. Wagner says, "without a long-enduring separation of colonists from their former species, the formation of a new race cannot succeed—selection, in fact, cannot take place. Unlimited crossing, unhindered sexual mingling of all individuals of a species will always produce uniformity, and drive varieties, whose characteristics have not been fixed throughout a series of generations, back to the primary form."

This sentence, in which Wagner himself comprises the main result of his investigations, he would be able to defend only if all organisms were of separate sexes, if every origin of new individuals were possible only by the mingling of male and female individuals. But this is by no means the case. Curiously enough, Wagner says nothing of the numerous hermaphrodites which, possessing both the sexual organs, are capable of self-fructification, and likewise nothing of the countless organisms which are not sexually differentiated.

Now, from the earliest times of the organic history of the earth, there have existed thousands of organic species (thousands of which still exist) in which no difference of sex whatever exists, and, in fact, in which no sexual propagation takes place, and which exclusively reproduce themselves in a non-sexual manner by division, budding, formation of spores, etc. All the great mass of Protista, the Monera, Amoebae, Myxomycetes, Rhizopoda, etc., in short, all the lower organisms which we shall have to enumerate in the domain of Protista, standing midway between the animal and vegetable kingdoms, propagate themselves *exclusively in a non-sexual manner*. And this domain comprises a class of organisms which is one of the richest in forms, nay, even in a certain respect the richest of all in forms, as all possible geometrical fundamental forms are represented in it. I allude to the wonderful class of the Rhizopoda, or Ray-streamers, to which the lime-shelled Acyttaria and the flint-shelled Radiolaria belong. (Compare chapter xvi.)

It is self-evident, therefore, that Wagner's theory is quite inapplicable to all these non-sexual organisms. Moreover, the same applies to all those hermaphrodites in which every individual possesses both male and female organs and is capable of self-fructification. This is the case, for instance, in the flat-worms, flukes, and tape-worms, further in the important Sackworms (Tunicates), the invertebrate relatives of the vertebrate animals, and in very many other organisms of different groups. Many of these species have arisen by natural selection, without a "crossing" of the originating species with its primary form having been possible.

As I have already shown in the eighth chapter, the origin of the two sexes, and consequently sexual propagation in general, must be considered as a process which began only in later periods of the organic history of the earth, being the result of differentiation or *division of labour*. The most ancient terrestrial organisms can have propagated themselves only in the simplest non-sexual manner. Even now all Protista, as well as all the countless forms of cells, which constitute the body of higher organisms, multiply themselves only by non-sexual generation. And yet there arise here "new species" by differentiation in consequence of natural selection.

But even if we were to take into consideration the animal and vegetable species with separate sexes, in this case too we should have to oppose Wagner's chief proposition, that "the *migration* of organisms and their formation of colonies is the *necessary condition of natural selection*." August Weismann, in his treatise on the "Influence of Isolation upon the Formation of Species,"[24] has already sufficiently refuted that proposition, and has shown that even in one and the same district one bi-sexual species may divide itself into several species by natural selection. In relation to this question, I must again call to mind the great influence which *division of labour, or differentiation*, possesses, being one of the necessary results of natural selection. All the different kinds of cells constituting the body of the higher organisms, the nerve cells, muscle cells, gland cells, etc., all these "good species," these "bonae species" of elementary organisms, have arisen solely by division of labour, in consequence of natural selection, although they not only never were locally isolated, but ever since their origin have always existed in the closest local relations one with another. Now, the same reasoning that applies to these elementary organisms, or "individuals of the first order," applies also to the many-celled organisms of a higher order which only at a later date have arisen as "good species" from among their fellows.

We are therefore of the same opinion as Darwin and Wallace, that the migration of organisms and their isolation in their new home is a very advantageous condition for the origin of new species; but we cannot admit, as Wagner asserts, that it is a *necessary* condition, and that without it no species can arise. Wagner sets up this opinion, "that migration is a necessary condition for natural selection," as a special "*law of migration*"; but we consider it sufficiently refuted by the above-mentioned facts. We have, moreover, already pointed out that in reality the origin of new species by natural selection is a *mathematical and logical necessity* which, without anything else, follows from the simple combination of three great facts. These three fundamental facts are—the Struggle for Life, the Adaptability, and the Hereditivity of organisms.

We cannot here enter into detail concerning the numerous interesting phenomena furnished by the geographical and topographical distribution of organic species, which are all wonderfully explained by the theory of selection and migration. For these I refer to the writings of Darwin,[1] Wallace,[36] and Moritz Wagner,[40] in which the important doctrine of the *limits of distribution*—seas, rivers, and mountains—is excellently discussed and illustrated by numerous examples. Only three other phenomena must be mentioned here on account of their special importance. First, the close relation of forms, that is, the striking "family likeness" existing between the characteristic local forms of every part of the globe, and their extinct fossil ancestors in the same part of the globe; secondly, the no less striking "family likeness" between the inhabitants of island groups and those of the neighbouring continent from which the islands were peopled; lastly and thirdly, the peculiar character presented in general by the flora and fauna of islands taken as a whole.

All these chorological facts given by Darwin, Wallace, and Wagner—especially the remarkable phenomena of the limited local fauna and flora, the relations of insular to continental inhabitants, the wide distribution of the so-called "cosmopolitan species," the close relationship of the local species of the present day with the extinct species of the same limited territory, the demonstrable radiation of every species from a single central point of creation—all these, and all other phenomena furnished to us by the geographical and the topographical distribution of organisms, are explained in a simple and thorough manner by the theory of selection and migration, while without it they are simply incomprehensible. Consequently, in the whole of this series of phenomena we find a new and weighty proof of the truth of the Theory of Descent.

END OF VOL. I.

www.ingramcontent.com/pod-product-compliance
Lightning Source LLC
Chambersburg PA
CBHW051344200326
41521CB00014B/2472